DIGITAL CONQUEST

Spatiality, Energy and Data Centres

Fanny Lopez and Cécile Diguet

BRISTOL
UNIVERSITY
PRESS

First published in Great Britain in 2025 by

Bristol University Press
University of Bristol
1–9 Old Park Hill
Bristol
BS2 8BB
UK
t: +44 (0)117 374 6645
e: bup-info@bristol.ac.uk

Details of international sales and distribution partners are available at bristoluniversitypress.co.uk

malaquais/ École nationale supérieure d'architecture Paris-Malaquais
LIAT Laboratoire Infrastructure Architecture Territoire

British Library Cataloguing in Publication Data
A catalogue record for this book is available from the British Library

ISBN 978-1-5292-5215-6 hardcover
ISBN 978-1-5292-5217-0 ePub
ISBN 978-1-5292-5218-7 ePdf

Cover design: Andy Ward
Front cover image: Jannes Linders
Bristol University Press uses environmentally responsible print partners.
Printed and bound in Great Britain by CPI Group (UK) Ltd, Croydon, CR0 4YY

FSC www.fsc.org MIX Paper | Supporting responsible forestry FSC® C013604

Bristol University Press' authorized representative in the European Union is:
Easy Access System Europe, Mustamäe tee 50, 10621 Tallinn, Estonia,
Email: gpsr.requests@easproject.com

Contents

List of Figures		v
About the Authors		viii
Acknowledgements		ix
Introduction		1
PART I	**The Smart City at Half-Mast: Storing Data to Counter Crises**	**11**
1	Modelling Degrowth: The Utopia of the Informational and Communicational City	13
2	The Internet Infrastructure: From Dispersion to Centralization	19
3	An Infrastructure without Architecture?	23
4	Scales, Typologies and Siting Strategies of Data Centres	29
PART II	**A Globalized Territorial Continuum**	**33**
5	In the Dense City	37
6	In Urban Outskirts, Growing Digital Areas	53
7	Isolated in Rural Zones	76
8	Growth, Acceleration and Digital Productivism	91
PART III	**Energy Disturbances and Conflicts of Use**	**95**
9	Rising Electricity Consumption and Territorial Destabilization	97
10	Île-de-France in the Quest for a Strategy	103
11	Infrastructural Redundancies and Difficult Pooling	109
12	The Dutch and Swedish Examples: Integrated Urban Approaches	120
PART IV	**Another Digital: Rethinking the Network and Its Infrastructures**	**133**
13	Cyberattacks and Political Intrusions: States Seeking Independence	135

14	The Peer-to-Peer Solution: Organizing a System Distributed among Equals	138
15	Self-Management and Technological Sovereignty: Decentralized Digital Infrastructures	144
16	Resilience and Inclusion: Systems Faced with Disconnection	155
17	The Ecological Footprint of Alternative Digital Infrastructures	162
PART V	**Alternative Futures: Three Possible Digital Worlds**	**167**
18	New Digital Worlds	169
Conclusion		178
Bibliography		181
Index		194

List of Figures

3.1	Entrance to the city of The Dalles	24
3.2	Google data centre, The Dalles	24
3.3	Entrance of the Google data centre, The Dalles	25
3.4	Amazon data centre, Port Morrow	25
5.1	32 Avenue of Americas, New York	38
5.2	60 Hudson Street, New York	39
5.3	Arrival of cables in 32 Avenue of the Americas	40
5.4	A conduit of cables	41
5.5	The circulation of flows under the roadbed	42
5.6	32 Avenue of the Americas, data cabinets	42
5.7	32 Avenue of the Americas, detail of a bay	43
5.8	32 Avenue of the Americas, safety cage	44
5.9	Antenna and crane on the roof of 32 Avenue of the Americas	45
5.10	Sabey tower, New York	46
5.11	111 8th Avenue in Chelsea	47
5.12	Orange data centre in Montsouris in Paris	48
5.13	Telehouse operator, boulevard Voltaire	49
5.14	AM4 (tower) and AM3 (horizontal building) of Equinix, Sciences Park	50
5.15	The data tower of Digital Realty, Sciences Park	51
5.16	Façade of the data tower, Sciences Park Amsterdam	52
6.1	Data centres in Santa Clara; only Digital Realty is recognizable	55
6.2	Vantage campus, Santa Clara; the structure of the new data centre being built	56
6.3	Coresite data centre, Santa Clara	57
6.4	Equinix data centre, San José	57
6.5	The OVH and Adobe data centres in Hillsboro	59
6.6	Stack data centre in Hillsboro surrounded by power lines with, in the distance, the substation	59
6.7	Stack centre in Hillsboro	60
6.8	Microsoft data centre, Grange Castle Business Park, 2020	63
6.9	Grange Castle plot undergoing transformation in Dublin	63

6.10	The Google data centre in Profile Park is one of the rare centres in the sector to distinguish itself with a mural on the façade, although this façade is practically invisible to the public	64
6.11	Data centre in the Schiphol Rijk zone	66
6.12	Data centre in the Schiphol Rijk zone	67
6.13a	Site buildings	71
6.13b	Site buildings	71
6.14	Plan of the site with the data centre plot	72
6.15	View of the project, 2018	73
6.16a	New Interxion data centre in La Courneuve	74
6.16b	New Interxion data centre in La Courneuve	74
7.1	Apple data centre in Prineville, Oregon	77
7.2	Facebook data centre in Prineville	77
7.3	Apple solar farm in Prineville	79
7.4	Amazon data centre in Umatilla	80
7.5	Part of the Microsoft data centre in Agriport A7	82
7.6	Microsoft substation in Agriport	83
7.7	Map showing Agriport, with its zones dedicated to data centres in red and greenhouses in yellow	84
7.8	The Google data centre in Agriport A7, 2019	85
7.9	First building of the Ersbo site, May 2020	88
7.10	Stackbo, Gävle site under construction, May 2020	89
7.11	The first building of the Tuna site in Sandviken, 2020	89
9.1	Share of communication technology of global electricity usage	98
9.2	Power lines in Prineville, 2017	100
10.1	Map of data centres in the Paris region, 2021	104
11.1	Roof of 111 8th Avenue	110
11.2	Access corridor for the data storage room, 111 8th Avenue, New York	111
11.3	Heat conduits, hot and cold water, 111 8th Avenue, New York	112
11.4	Fuel oil tank and heat storage	113
11.5	Portland Dispatchable Standby Generators, photos of backup generators at the Via West data centres and other buildings connected to the micro-grid	115
11.6	Data centre of the city of Paris, 2022	118
11.7	Céleste offices and data centre in Champs-sur-Marne	118
12.1	Low, medium and high scenarios in MW	123
12.2	Electricity availabilities in the substations	124
12.3	'Substation' buildings	125
12.4	Former printing plant transformed into data centres in Kista	128
12.5	Stack of the Fortum plant, urban heating system in Kista	129
15.1	Storage architecture at Aquilenet	148

15.2 Homepage of the Anarchaserver site 153
16.1 Repair of the meshed Wi-Fi network node in Red Hook after Hurricane Sandy 156
16.2 Sign advertising the 'Where's the B61 Bus?' Web application available through the local Red Hook Wi-Fi network 157
16.3 Infrastructural mediation map 159

About the Authors

Cécile Diguet is Head of studio dégel. She has 20 years' experience in urban planning and local public policies from large metropolis master planning like Tehran, Amman, Paris or Medellin, to the urban design scales of projects in Greater Paris, Buenos Aires, Tbilissi and Rio de Janeiro. She has been working relentlessly in interweaving sustainable planning and care for social uses at all scales, to design for urban resilience, heritage protection and planet livability. She has been involved in recent years in thinking of a better way to integrate and minimize the impact of digital infrastructures on cities and territories. She regularly curates exhibitions on architecture, urban design and landscape. Cécile also teaches at Sciences Po Paris on cities and digital infrastructures.

Fanny Lopez is a historian of architecture and technology, and Full Professor at the School of Architecture Paris-Malaquas at the Paris Sciences et Lettres University. She is co-director of the LIAT (Laboratory Infrastructure, Architecture, Territoire). Her research and teaching activities focus on the spatial, territorial and environmental impact of electricity and digital infrastructures. Her publications include: *Le rêve d'une déconnexion. De la maison autonome à la cité auto-énergétique* (La Villette, 2014), translated in The Dream of Disconnection. From the Autonomous House to the Self-Energy Territories (Manchester University Press, 2021); *L'ordre électrique, infrastructures énergétiques et territoires* (Métis Presses, 2019); *Local Energy Autonomy: Spaces, Scales, Politics* (Wiley, 2019, with O. Coutard and M. Pellegrino), and *À bout de flux* (Ed. Divergences, 2022). In 2021, she reviewed the Prize of the Association for the History & Sociology of Energy.

Acknowledgements

Thanks to Eileen Powis for the translation of the French version of the book *Sous le feu numérique, spatialités et énergie des data-centers* published by MētisPresses in 2023.

Many thanks to Maximilian Gawlik for the research we carried out together in Dublin, Stockholm and Amsterdam and for sharing these photos.

Thanks also to all those who took the time to share precious documents and archives, to exchange ideas and/or to accompany visits to the sites.

We would like to thank the Agence de l'environnement et de la maîtrise de l'énergie (ADEME) and the Caisse des Dépôts et Consignations for their support, which made our study trips to the United States and Europe possible.

Introduction

In April 2021, Amazon Web Services France was denied authorization to build a data centre of 33,100 m^2 in Bretigny-sur-Orge in the northern part of the Essonne department. It was a first in France. The order signed by the prefect of the Île-de-France specified that the request to build this complex for use as a data storage centre was incompatible with the orientations of the regional development policy, in particular with the master plan of the Île-de-France region and the regional plan for climate, air and energy. According to this same source,[1] the Amazon project did not include any analysis of vacancy or the densification potential of the existing urban fabric so that the necessity of artificializing this 9.7-hectare space could be evaluated. It was moreover noted that no solution for recycling waste heat was proposed by the digital industrialist. Over the last ten years, the monopolizing of real estate and energy by large private digital companies has become a major concern for the public authorities that are confronted with conflicts of use. In Île-de-France, there are 168 data centres for 115[2] sites and an electricity consumption that exceeds 1 GW (1 GW = 1,000 MW), the equivalent of the electricity production of the two pressurized-water reactors of the nuclear power plant in Nogent-sur-Seine. Buoyed by the digitalization of the economy, the explosion of data exchanges, the Cloud and the connected Internet of Things (IoT), digital infrastructures will be one of the most important areas of electricity consumption in the 21st century. A new stage in network urban planning, the digital city was often analyzed in terms of uses and practices, services and events, considering the spatiality and energy impact of its infrastructures as secondary. Some 15 years after the emergence of the smart city concept created by industrialists and the Cisco and IBM engineers, during the financial crisis affecting the United States, economic and urban forecasting actively worked on setting up this global infrastructural product. Private companies approached the historic urban

1 Order no idf-2021-04-12-00012.

2 This figure resulted from counting work conducted in 2023 with the Institut Paris Region in the framework of its study on data centre development in Île-de-France.

operators to attempt to bring together, store, sort and analyze the data that made it possible to maintain and optimize the city of flows, and to counter the crises – technical, ideological, climate and energy – that strike it. The fascination that accompanied this digital urbanity and its economy sometimes forgot its materiality. Paradoxically, in the dematerialization imagination that is part of the cybernetic utopia that made the smart city a sort of magical territory in which flows circulated without any physicality, nothing is a greater consumer of space and energy than the digital infrastructures. The most pessimistic projections for 2030 reach a consumption maximum of 51 per cent of worldwide electricity dedicated to the IT sector in its totality.[3]

The smart city is not solely responsible for the growth of the digital footprint, with a host of other uses contributing to it, including the continuing growth of digital content (videos, online advertising, applications, etc) and increasingly energy-intensive calculation and data processing tools (AI, etc).

In the worldwide interconnection and storage system, the United States and Europe hold a central place in which, for the latter, the top four hubs are London, Frankfurt, Amsterdam and Paris. The euphoria of the French market was accelerated over the last few years by a unique phenomenon on the European scale: the structuring of a double hub of worldwide scope on the national territory. According to TeleGeographic, a company that studies the market of digital flows,[4] Paris and Marseille are now in the top 10 worldwide cities that have the greatest digital flow. In 2021, the city of Marseille went from 122nd position to 9th, thus surpassing Hong Kong and New York. This industry's growth is continuing, further strengthened in 2021 by the creation of new optical fibre underwater cables (the Peace, 2Africa, Ellalink, Maréa, Dunan, Amitié cables), but also by the evolutions of the technologies deployed themselves (optronics, wavelengths, etc) that strengthen the data market.

Will the digital make it possible to economize energy? As soon as 2016, according to the ADEME report by the Eco-Info research group of the Inria, most of the studies on this subject are not sufficiently documented or serious.[5] A paradox emerges: the corollary of the exponential growth principle of the data required for the digital industry to function is an increase in space and energy needs. The digital city partially contains the problem that it claims to solve. In an intermingling of technological, societal

[3] Anders S. G. Andrae and Edler Tomas, 'On Global Electricity Usage of Communication Technology: Trends to 2030', *Challenges*, 6, 2015, pp 117–57.

[4] Exchanges per second.

[5] Deloitte Développement Durable, Eco Info, Futuribles and the CRÉDOC, *Potentiel de contribution du numérique à la réduction des impacts environnementaux. État des lieux et enjeux pour la prospective*, ADEME report, 2016.

and power systems, the digital era started with the advent of the networked society[6] a new territorialization of flows. Our hypothesis is that in order to spatialize, evaluate and measure the energy impact of the digital technical choices on the territories, it is necessary to analyze the siting and architecture of digital infrastructures, in particular data centres that are data storage and interconnection places, but also the electrical infrastructure that makes them function. How can this reticular complex be organized? Is it possible to better integrate these structures into urban and rural territories, and what about energy pooling like heat recovery and the backup infrastructure sharing with which data centres are equipped? Can this unthought-of spatial dimension become a more virtuous vector of development, a more virtuous energy and exchange producer? Between the ultra-centralizing spatiality of hyperscale data centres and the explosion of micro-data centres needed to develop the IoT and edge computing, what are the architectural and urban integration prospects of data centres? At a time when we are beginning to evaluate the dizzying increase in the energy footprint of digital infrastructures, what pooling can be organized and according to what governance?

This book mobilizes the methodological and analytical frameworks of infrastructure studies, urban studies and the sociology of technologies and innovations. The specificity of the Digital Technical System (DTS) compared to other Large Technical Systems[7] is documented in the histories of techniques, but its infrastructural-spatial and energy dimension still remains underestimated. In the field of social sciences and development disciplines, the state of knowledge relative to the smart city as an evolution of the urbanism of networks was largely analyzed as a new urban ontology.[8] Several researchers analyzed the evolution of telecommunication networks from the socio-spatial perspective that made it possible to better understand their link to metropolitan developments.[9] The complexity of the development

6 Manuel Castells, *La Société en réseaux. L'ère de l'information*, Paris, Fayard, 1998.

7 Thomas Parke Hughes, *Networks of Power: Electrification in Western Society, 1880–1930*, Baltimore, MD, Johns Hopkins University Press, 1983; Thomas Parke Hughes and Renate Mayntz, *The Development of Large Technical Systems*, Frankfurt, Campus, 1988; Wiebe E. Bijker, Thomas Parke Hughes and Trevor Pinch (eds), *The Social Construction of Technological Systems*, Cambridge MA, MIT Press, 1984.

8 Anthony M. Townsend, *Smart Cities, Big Data, Civic Hackers, and the Quest for a New Utopia*, New York, W. W. Norton & Co, 2013; Antoine Picon, *Smart cities: Théorie et critique d'un idéal auto-réalisateur*, Paris, B2, 2013; Antoine Picon, *Smart Cities – A Spatialised Intelligence*, Chichester, Wiley, 2015; Ali Fard and Taraneh Meshkani (eds), *Geographies of Information*, New Geographies 07, Harvard University Graduate School of Design, October 2015.

9 Stephen Graham, 'Global grids of glass: on global cities, telecommunications and planetary urban networks', *Urban Studies*, 36, 1999, pp 929–49; Stephen Graham and Simon Marvin, *Splintering Urbanism: Networked Infrastructures, Technological Mobilities and the Urban Condition*, London, Routledge, 2001; Jonathan Rutherford, *A Tale of Two Global Cities: Comparing the Territorialities of Telecommunications Developments in Paris and*

of metropolitan telecommunications involved a group of contextual and historic processes, interests and practices on several spatial scales,[10] made more complex by the emergence of the Internet.[11] It is from the angle of the urban externalities of the digital economy that the spatial effects were most often discussed: practice of the public space in its informational, communicational and event-oriented, but also domestic dimension (from the smart counter to the connected vehicle). Most of the socio-spatial studies endeavour to observe the new uses and practices that the digital city offers in terms of services (e-mobility,[12] e-commerce, collaborative economy, etc). The territorial dimension has also been questioned through the idea of digital territorialities[13] and the way in which the DTS, by eluding scales, modifies our understanding of the fundamental qualities of space.[14] The infrastructural and energy, in particular electrical, aspect remains, however, in the background despite a socio-technical corpus that has analyzed, these last several years, the infrastructure crisis and the matrix reorganizations that are emerging.[15]

In 2018, when we began the investigation, there were very few publications examining the spatiality of data centres, their environmental impact and the energy policies associated with their development.[16] In just a few years,

---

London, Aldershot, Ashgate, 2004; Jonathan Rutherford, 'Networks in cities, cities in networks: territory and globalisation intertwined in telecommunications infrastructure development in Europe', *Urban Studies*, 42(13), 2005, pp 2389–406.

10 Jonathan Rutherford, 'Le déploiement des réseaux de télécommunications dans les métropoles européennes: l'emprise des territorialités Multiscalaires', *Métropolis Flux*, 75, 2009, pp 6–15.

11 Daniel Schiller, *Digital Capitalism: Networking the Global Market System*, London, MIT Press, 1999; Gabriel Dupuy, *Internet: Géographie d'un Réseau*, Paris, Ellipses, 2002.

12 Dominique Rouillard and Alain Guiheux, *Door to door: futur du véhicule, futur urbain*, Paris, Archibooks, 2016; and *Le projet Hub – L'architecture des nouvelles mobilités*, Geneva, Métis Presses, 2020.

13 Pierre Musso, 'Critique de la notion de "territoires numériques"', *Quaderni*, no 66, spring 2008, pp 15–29.

14 Asta Vonderau, 'Scaling the cloud: making state and infrastructure in Sweden', *Ethnos*, 84(4), 2018, pp 698–718; Soline Nivet, *Architecture et numérique: Infrastructures, Interfaces & Institutions d'un 'écosystème' parisien, HDR*, Université Paris-Est Sup, 15 December 2020. This work concerns the activities of Free, one of the principal French Internet Service Providers. The unpublished HDR report analyzes the spaces dedicated to optical fibre, 4G telephony and digital data storage in Paris between 2007 and 2017.

15 See, on the network crisis, Gabriel Dupuy, 'Fracture et dépendance: l'enfer des réseaux?', *Flux*, 1(83), 2011, pp 6–23; Cédric Le Bris and Olivier Coutard, 'Les réseaux rattrapés par l'environnement? Développement durable et transformations de l'organisation des services urbains', *Flux*, 4(74), 2008, pp 6–10; Fanny Lopez, *L'ordre électrique, infrastructures énergétiques et territoires*, Geneva, Métis Presses, 2019.

16 See in particular: Paul N. Edwards, *Computer Models, Climate Data, and the Politics of Global Warming*, Cambridge, MA, MIT Press, 2013; Mél Hogan, 'Data flows and water

the number of studies has increased significantly, placing the objects of the digital technical system at the centre of science and technology studies, at the disciplinary intersection of media and communication studies, environmental studies and public energy policies. In order to move away from digital immediacy and contemporaneity, a number of studies have focused on the relationship with time, placing case studies in a long-term context, as in the special issue of *New Media and Society* on 'Infrastructural Temporalities of Data Centres', edited in 2023 by Julia Velkova and Jean-Christophe Plantin.[17] The authors take a cultural approach to analyzing contemporary Internet infrastructures through their temporalities and multiple mediations. This is also the case with Jenna Burrell's[18] research on the Facebook data centre in Prineville, Oregon. She shifts the analysis of the facility over time, delivering an environmental and energy history of the region that shows that the ground has long been prepared for data centres through legislation and zoning, and the deployment of electricity and fibre optic infrastructures, etc. The development of data centres is nothing more than a new stage in a project to develop and transform the territory for capitalist ends. This approach is similar to that of Quincy Childs on data centres in the north of Sweden, or of Patrick Brodie and Sebastián Lehuedé who, by placing data centres in the technological history of capitalism, and of these protests, insist on the social and environmental injustices they reproduce, thus extending Anil Agarwal and Sunita Narain's[19] concept of environmental colonialism. From a perspective that concerns energy and spatial as much as political issues, the collective work *Data Centers: Edges of a Wired Nation*, edited by Monika Dommann, Hannes Rickli and Max Stadler,[20] has provided enlightenment on the impact of the Swiss digital industry by noting the interweaving of narratives on technical-national aspirations, the deregulation of telecommunications companies, the heritage of the Cold War, data security commerce and the interdependence between Switzerland and China.

This work relies on 20 or so case studies in the field. The spatial and energy planning strategies of data centres are documented and analyzed: they are

woes: The Utah data center', *Big Data and Society*, 2(2), 2015, pp 1–12; M. Hogan, 'The data centre industrial complex', in M. Jue and R. Ruiz (eds), *Saturation: An Elemental Politics*, Durham, NC, Duke University Press, 2021, pp 283–305.

17 Jean-Christophe Plantin and Julia Velkova, 'Data centers and the infrastructural temporalities of digital media: An introduction', *New Media and Society*, 25(2), 2023, pp 273–86.

18 Jenna Burrell, 'On half-built assemblages: waiting for a data center in Prineville, Oregon', *Engaging Science, Technology, and Society*, 6, 2020, pp 283–305.

19 Anil Agarwal and Sunita Narain, *Global Warming in an Unequal World: A Case of Environmental Colonialism*, New Delhi, Centre for Science and Environment, 1991.

20 Monika Dommann, Hannes Rickli and Max Stadler (eds), *Data Centers: Edges of a Wired Nation*, Zurich, Lars Müller, 2020.

the machine-buildings in the heart of Paris or New York; the metropolitan clusters of Amsterdam, Dublin and Stockholm; the digital outskirts of Portland, Silicon Valley and Paris; the transformation of the rural territories of Prineville in Oregon and Gävle in northern Sweden. Derived from three years of research in the United States and Europe (Spain, France, Ireland, Sweden and the Netherlands), the sources of this text come from field visits and over 100 interviews with the principal actors of electricity companies and urban services of the territories concerned. Part of the raw material comes from interviews and site visits conducted in the framework of research for ADEME in 2018.[21] But most of it comes from later research carried out between 2019 and 2021, and interviews with European electricity distribution and transmission operators.[22]

Using a socio-technical approach blending the history of urbanism and that of electricity, this text proposes questioning, in three parts, the territories transformed by the electrical-digital condition. Part I, taking a detour vis-à-vis the large-scale experiments of the smart city, heavily covered by the media, proposes studying the territorial transformations generated by the storage infrastructures, supports of contemporary digital life. The dominant narrative of the smart city today is one of an approach that is almost entirely based on solving complex urban problems on information and communication technologies: mobilities, air pollution, water or electricity network management. This technical and computational approach also seems to be one of the principal tools of the neo-liberalization of the cities,[23] through the shift of the control of the urban data of the public authorities to private digital companies. How can the ambiguity of the path that is emerging, both narrative, project and operating principle, which constantly pushes back the limits of growth by blurring the boundaries between public and private, liberty and control, be noted? The future promised by the digitalization of the world, cities and territories, does not seem to reorient

21 Part of the results of the research were presented in an institutional report: Cécile Diguet and Fanny Lopez, *L'impact spatial et énergétique des data centers sur les territoires*, ADEME report, 2019. This report was divided into seven parts: The digital infrastructure – a growing footprint on the territories; Territories of the digital – concentration and dispersion; Interplay of actors and governance of the digital territories; The data centres, new pieces of the energy puzzle; Public, citizen initiatives and alternative digital infrastructures; Integrated approaches; and An energy, digital, territories forecasting. If it relies partially on this work, the text presented here is different and enriched by later original research conducted in 2020 and 2021.

22 In particular: RTE, Enedis, ESB, EirGrid, Tennet, Alliander, Liander, Ellevio, Vattenfal and Svenska Kraftnät.

23 Evgeny Morozov and Francesca Bria, *Rethinking the Smart city, Democratizing Urban Technology*, New York, Rosa Luxemburg Stiftung, 2018.

the global 'capitalocene'[24] trend towards an adaptation of humans' ecological footprint to the load capacity of the planet.[25] 'The digital functions with growth', data centre operators constantly repeat.[26] How can urban and spatial narratives be encouraged to emerge that are alternatives to that of the smart city[27] or the globalized city of GAFAM?[28] Can we envisage the links between the digital, the city and the territory differently? Our conviction is that the structuring of urban and spatial narratives that are alternatives to those of the smart city[29] or the globalized digital city occurs above all by an infrastructural reorganization.

Part II concerns the spatial form of the digital socio-technical system. It wishes to make a connection between the architecture of information systems (software and hardware) and the architecture of the infrastructures that serve them. Data centres are the central infrastructure of the digital technical system; they are the network's nodes. A data centre is defined as a building for data hosting and interconnection. A data centre can contain different technologies depending on application needs, for example, computing servers for high-performance computing centres, storage bays or network equipment for telecom operator centres (like the meet-me rooms in which all the telecom operators are connected to one another). It is equipped with cooling and heat recovery systems as well as backup equipment: batteries, undulators and generators. Far from the visible and monumental architecture of certain meccas of the network like the head offices of large digital companies like Apple (architect Norman Forster), Amazon (NBBJ), Google (Heatherwick Studio and Bjarke Ingels), Facebook/Metaverse (architect Frank Gehry) and Microsoft (GF, LMN Architects, WRNS Studio and NBBJ), data centres have long remained more discreet in their architectures. Digital constructions are similar to those of shopping centres and logistics warehouses, they borrow from different registers, long lost in the outskirts saturated and drowned in the mass of box architectures. They offer, however, a broad typological range: anonymous, standardized and flexible 'boxes'; manifesto architectures on high-tech campuses; transformed buildings such as former offices, telephone exchanges, industrial sites and obsolescent

24 Jason W. Moore (ed), *Anthropocene or Capitalocene?: Nature, History, and the Crisis of Capitalism*, Kairos, PM Press, 2016.

25 Tim Jackson, *Prosperity without growth: Foundations for the Economy of Tomorrow*, London and New York, Earthscan/Routledge, 2009.

26 Interview with Fabrice Cocquio, President of Interxion France, Paris, September 2021.

27 Valérie Peugeot, 'Collaborative ou intelligente? La ville entre deux imaginaires', in Maryse Carmes and Jean-Max Noyerp (eds), *Devenirs urbains*, Paris, Presses des Mines, 2016, pp 43–64.

28 GAFAM refers to Big Tech: Google, Amazon, Facebook, Apple, Microsoft.

29 Peugeot, 'Collaborative ou intelligente?'

commercial zones. In a boomerang effect, this storage infrastructure layer impacts every territory: urban, suburban and rural. Visible or discreet, even hidden underground or in bunkers to avoid the climate or terrorist threat, data centres are present everywhere in the city centre, in urban outskirts and in remote and deserted regions, provided there is a good electricity connection nearby. Whether they saturate electricity grids, whether they are integrated into heat exchange or electricity circles of variable perimeters (block, neighbourhood, city, territory) or they attempt electricity autonomy, they redefine each time the energy project of the territories in which they are installed. Why is there so little architecture for this infrastructure?

Part III is focused on the impacts of the digital infrastructure on the electricity grids and the attempts at pooling. Is it possible to envisage an integration that is the most eco-systemic of the technical systems of these objects (energy, water, backup infrastructures) whose governance and regulation modalities elude for the moment the public actors of electricity and urban planning? What control and what hybridization of uses can be imagined? The diversification of the major digital actors (Big Tech) towards urban services is not completely new, but the convergence of their monopolistic position on the collection and storage of personal data, on software, commerce and communication, on the media and culture, that is, all the exchanges of goods and information, from contents to distribution networks and storage spaces, shows a new infrastructural appetite that is expressed by a replacement of the historic operators. The complex relations they have with electricity operators destabilized by the constant increases in demand for an electrical connection have forced many metropolises to regulate, even prohibit, the installation of data centres outside reserved areas, following the example of Amsterdam and Stockholm. Everywhere in Europe and the United States, privileged terrains for this study, electricity transmission and distribution grids require strengthening to face the growth of their electricity consumption. In many cities, as substations are saturated, conflicts of use have appeared, sometimes threatening the grids' equilibrium.

The technical-critical approach to the digital mostly follows two lines of analysis. The first is that of data confidentiality and respect for private life, endangered by the development and use of predictive systems for surveillance and population control purposes[30] and for economic interests with advertising profiling. The second line is that of digital restraint,[31] which

---

30 Sarah Brayne, *Predict and Surveil: Data, Discretion, and the Future of Policing*, New York, Oxford University Press, 2021.

31 Françoise Berthoud and Laurent Lefevre, 'Impact écologique du Big Data', in Mokrane Bouzeghoub and Rémy Mosseri (eds), *Les Big Data à découvert*, Paris, CNRS, 2017; Françoise Berthoud, Amélie Bohas, Laurent Lefevre and Marianne Parry, 'Transformation numérique et transformation énergétique – Digital transformation and energy transformation', in Agathe Euzen, Stéphanie Thiébault and Bettina Laville, *Quelles*

aims at reducing the use of electronic services and devices that require large quantities of raw materials, the extraction of precious ores being necessary for their functioning. The energy question is increasingly present, as shown by the publication of the journalist Guillaume Pitron,[32] but also certain reports like those of The Shift Project.[33] New data storage buildings are generally energy-efficient. It is the Power Usage Effectiveness (PUE) that sets the sector's energy efficiency. Without being a sufficient indicator, it is universally recognized. It is the ratio between the electricity consumed by the servers and that which enters the building. The French average PUE is around 1.36,[34] which means that for 1 watt consumed by information technology, 1.36 watts must enter the data centre.

One of the specificities of this research is to bring the digital network and the electricity grid closer together, to analyze their complex and subtle intertwining in the territories. The large digital companies constantly extend their private networks using the channels and resources of public electricity infrastructures needed for them to function and grow, whereas the consequences on the territories cause strong tension, even electricity shortages. In the United States as in Europe, MW conflicts of use destabilize the urban and energy planning of many municipalities largely left behind by this phenomenon. For the electricity operators, the situation is just as critical, the construction of public distribution infrastructure, like the substations, has a difficult time following the rhythm of the data sector, whose demands are ever-increasing. However, no national regulations have today ruled on the capitalistic power of the digital giants. On the contrary, the current regulations even permit them to roll out their data centre on any available land for sale and when the timeframe for building infrastructures does not coincide with their growth plan, they pay for all or part of their connection infrastructure themselves and sometimes their production infrastructure, forming strange alliances with the electricity companies.

---

solutions pour le changement climatique? Which solutions for climate change?, Paris, CNRS, 2015; Anne-Cécile Orgerie and Laurent Lefèvre, 'Energy-efficient data transfers in large-scale distributed systems', in Ishfaq Ahmad and Sanjay Ranka (eds), *Handbook of Energy-Aware and Green Computing*, Chapman & Hall, 2011; Fabrice Flipo, *Le numérique, une catastrophe écologique*, Paris, L'Échappée, 2021; Fabrice Flipo, *L'impératif de la sobriété numérique: Les enjeux des modes de vie*, Paris, Matériologiques, 2020; Fabrice Flipo, *La face cachée du numérique*, Paris, L'Échappée, 2013; François Deltour, Michelle Dobré, Fabrice Flipo and Marion Michot, *Peut-on croire aux TIC vertes? Technologies de l'information et crise environnementale*, Paris, Presses des Mines, 2012.

32 Guillaume Pitron, *L'enfer numérique: Voyage au bout d'un Like*, Paris, Liens qui libèrent, 2021.

33 Hugues Ferreboeuf (ed), *Lean ICT, Pour une sobriété numérique*, The Shift Project report, October 2018.

34 From the website Cloudscience, consulted on 20 March 2024.

Part IV explores the angles through which the current digital technical system could be reformed, from both an environmental and a democratic perspective. Digital activists defend a free and neutral Internet, in which the operating rules are collectively and democratically decided, without surveillance and capture of personal data. In France, associations like the Quadrature du net, the FFDN and Framasoft, working with their European counterparts, act on this issue. They are the heirs of pioneering associations that, from the 1990s, specialized in the defense of fundamental rights on the Internet, such as the Associations des utilisateurs d'Internet and the IRIS (Imaginons un Réseau Internet Solidaire). They propose software and free tools, independently of any commercial interest and far from the famous 'if it's free, you're the product', all evoking the free services of the major digital actors like Google and Meta, which in exchange collect and sell Internet users' data. Félix Tréguer, researcher and former member of the Quadrature du net, traces the history of these actors in a recent work[35] that brings to light the major difficulties of this alternative world faced with the very powerful economic actors of the digital.

Several dimensions of digital alternatives, but also of technological sovereignty, are explored here. They are incarnated in the technical and governance systems anchored in different cultures and strategies, with sometimes divergent objectives. What are the ambivalences and challenges in taking over the digital technologies and infrastructures between civil societies and states? How can the spatial and environmental effects of a larger deployment of alternatives be evaluated in terms of networks and storage? What do isolated micro-societies in India that need to connect to the global Internet to do business have in common with vulnerable neighbourhoods of world cities like New York that must ensure the continuity of connections despite hurricanes? And what are the commonalities between censured political activists in authoritarian countries and initiatives that aim, above all, in well-connected societies, to encourage technical autonomy and the reappropriation of central tools of daily life? The question here is to better understand the links connecting technical proposals, objectives and governance. The countries are part of a centralized system that they attempt to increasingly control for internal political reasons and geopolitical motifs. The question of sovereignty constitutes above all a defensive and protectionist mechanism. Civil society guides Internet architecture towards a more decentralized system, one that is even more widely distributed, with associative, cooperative and local actors who work on relocating Internet benefits. Technological sovereignty then proposes a reappropriation that emancipates a digital system that can recover its shared nature.

[35] Félix Tréguer, *L'Utopie déchue, une contre-histoire d'internet, XVe–XXIe siècle*, Paris, Fayard, 2019.

PART I

The Smart City at Half-Mast: Storing Data to Counter Crises

Despite the contribution of science fiction at the beginning of the century that had anticipated a number of its abuses and promises, notably on the use of personal data and the limits of prediction, the conceptual success of the smart city marked the 2000s. Both a technical imaginary dimension and a group of connected objects, this global infrastructural product, created by industrialists and the Cisco and IBM engineers, then relayed by the information technology giants, dominated economic and urban forecasting. The dream of order was no longer achieved by a plan created by the urbanist-designer, but by the predictability of engineers from the IT sector whose programs made it possible to imagine homeostatic and self-referential systems. Their determination to incarnate this urban performance ideal through a new neighbourhood, a new head office or a new city, seems, however, to have raised their claim to compete with the local communities a notch, with the platformization of urban management towards which the data converge by means of embedded sensors in our mobile phones and connected objects.

But ten years after the first international Smarter Cities Challenge,[1] in spring 2020, the failure of one of the most celebrated smart city projects of

[1] IBM made its consultants available to regional administration actors to imagine 'smarter' cities. Over 2,000 selected local administrations consequently shared their data with this tech giant and in return benefitted from their strategic planning expertise through data management on a chosen sector: transportation, energy, water management, social services, security, etc. Buoyed by the explosion of the digitalization of the economy and data exchanges, the program continued in 2011, with the marketing of the Intelligent Operation Center for Smarter Cities, which proposed infrastructures and services to public actors and urban planning structures. Anthony M. Townsend, *Smart Cities, Big Data, Civic Hackers, and the Quest for a New Utopia*, New York, W. W. Norton & Co, 2013; Antoine Picon, *Smart Cities A Spatialised Intelligence*, Chichester, Wiley, 2015.

Sidewalk Labs (a subsidiary of Alphabet, Google's parent company), launched in 2017 in Toronto, marked a turning point. In the southwest part of the city centre near the port zone, the optimized urban services (smart bench, delivery robot, heated bike lane, etc) of 5 hectares of Quayside, integrated into the 77 hectares of the Innovative Design and Economic Acceleration (IDEA) District, would come to naught. Officially, Google pointed to the new economic uncertainties linked to the Covid pandemic, but the impact of the citizen opposition movement must not be underestimated in this about-face. Since 2018, #BlockSidewalk mobilizations have continuously denounced the appropriation of personal data for commercial purposes, private urbanism and the privatization of public resources, as well as the urbanistic and real estate weakness of the project. Opposition was further strengthened by the Facebook-Cambridge Analytica scandal that brought to light the issue of the confidentiality of data and respect for private life.

The project of a governance by the organizing power of algorithms and numbers is an old one as was shown by Alain Supiot, anthropologist and philosopher of the law. From antiquity to the Renaissance, the imaginary dimension of an ideal order that is based on mathematics has continuously fascinated lawmakers. 'The project of a society made transparent to those who direct it is found today in information technology tools, the means of implementation that neither Bodin, nor the Chinese jurists would have dared to dream about.'[2] This change appears as the symptom of deep institutional crises encountered by our society. Alain Supiot asserts that 'the law, democracy, the state and all the legal frameworks that we continue to refer to, are radically changed by the resurgence of the old Western dream of a harmony founded on computing'.[3] A new stage of Foucault's biopolitics, the digital permits useful computing to reach a new level of precision, tracked and in real time, to better administer and govern, surveil and punish. Everything is henceforth guided by the paradoxical hope of containing or controlling the crises, the climate and the environment through a new and almost infinite number of predictive data.

[2] Alain Supiot, *La gouvernance par les nombres*, Course at the Collège de France (2012–14), Fayard, 2015, p 134. See in particular the chapters 'Le rêve de l'harmonie par le calcul' and 'L'essor des usages normatifs', pp 103–55.

[3] Supiot, *La gouvernance par les nombres*, p 105.

1

Modelling Degrowth: The Utopia of the Informational and Communicational City

There is a correlation between the economic and climate crisis of the late 2000s and the success of the smart city service offering that made the global triumph over the local. Intelligence is in the machine that supposedly has the capacities to predict and treat large-scale problems, a capacity that the human being does not have to measure the effects on the small scale, her or his city, commune, neighbourhood. For the historian of sciences Orit Halpern, 'the preoccupations concerning climate change, the increasing scarcity of energy sources and the collapse of security and the economy make urbanists, investors and governments turn to smart infrastructures envisaged as a value production place and as a possible salvation for a world ceaselessly defined by disasters and crises'.[1] What is the future promised by the smart city? Is the scenario that of growth completed in a controlled manner[2] by the adaptation of human beings' ecological footprint to the planet's load capacity? At the start of the 2010s, the caricatural enthusiasms of Jeremy Rifkin on the digital, post-carbon productivism and green growth remained discreet on the question of the energy impact and the effects of the digital on the finitude of resources. The scenarios presented by the promotors of the smart city seem to be part of a perspective that aims at pushing back crisis situations (economic, environmental, political) by better management and anticipation of consumptions and current production (notably through the growth of renewable energies), without, however, the depletion of

[1] Orit Halpern, 'L'architecture comme machine: la ville intelligente déconstruite', in Andrew Goodhouse (ed), *Quand le numérique marque-t-il l'architecture?*, London, Stenberg Press, 2017, p 126.

[2] Jackson, 2009.

resources (ores, arable land, sand, drinking water, etc) and various types of pollution being a lever likely to more globally reorient our impact on the 'earth system', as a certain Richard Buckminster Fuller liked to call it. In his World Game, Fuller proposed a logistical vision of the network that was based on a societal and environmental reflection. About 20 years before the warning issued by the Club of Rome, Fuller invented a participatory smart network or a network game that brought together on a worldwide scale a catalogue of resources (natural, human); data (meteorological, climate, production); needs, etc. From an environmental and humanistic viewpoint, echoing the information science utopia of a peer-to-peer Internet, the players could predict and guide regulations and the fair division of this ensemble available on the earth scale. Beyond the pedagogic character, the effects of this tool on the reel and governance remained difficult to measure, but were a promise of a utopia that allied technology, ecology and the sharing of resources and therefore of wealth.[3] With the World Game, Fuller thought of the first smart network. Premonitory and enthusiastic, he anticipated that it would not be necessary to go to Russia to consume the electricity produced by its powerful hydroelectric dams. Fuller developed the World Game in the 1960s as a management tool for the planet's stock of energy. It concerned making what exists optimizable, and creating a sharing system in real time at more local scales. From a counterculture perspective that brought together the distributed character of the first age of the Web and a more globalist vision of knowledge and resource-sharing, Fuller described an Internet utopia, running counter to the performative cybernetic visions of Norbert Wiener or the centralizing and vertical imaginary dimension of Mumford's 'mega-machine'.[4]

From the Greek *kubernesis*, 'the art of guiding, governing', cybernetics explores the command, modelling and regulation processes of exchanges based on information and data.[5] This science was notably defined by the mathematician Norbert Wiener, in 1948, in *Cybernetics: Or Control and Communication in the Animal and the Machine*, in which he presented the parallel between the functioning of individuals and machines and how society could be reorganized based on this programmatic functioning and the machines' capacity to be able to bring together, sort and store information to make it usable. Adaptive behaviour characterizes the technological condition

3 Alain Gras, *Grandeur et dépendance, sociologie des macro-systèmes techniques*, Paris, Presses Universitaires de France, 2013; Jacques Ellul, *Le système technicien,* Paris, Calmann-Lévy, 1977.

4 The idea of the mega-machine was introduced by Lewis Mumford in *Technics and Civilization*, New York, Harcourt, Brace & Co, 1934.

5 Steve J. Heims, *Constructing a Social Science for Postwar America*, Cambridge, MA, MIT Press, 1991; Gabriel Dupuy, *Aux origines des sciences cognitives*, Paris, La Découverte, 1994.

of the cybernetic paradigm. The ancestor of the smart city, or rather its toolbox, thrives on the threat of a notably urban crisis that it claims to control and dominate, partially by adapting and creating a reactive environment since the 1960s. The cybernetic and computer technology utopia was structured in a period marked by the end of the modern movement. It was the architect-information specialist Nicholas Negroponte, founder of the Architecture Machine Group at MIT, who proposed to 'save the city from its modern ruins'.[6] He was the inventor of the idea of a 'reactive environment', the dream of an augmented, adaptable architecture that anticipated and withstood deterioration, and which restructured itself in the event of alteration. In a context of urban crisis and strong racial segregation, the ideas of Forrester and Negroponte[7] on obsolescence and urban degeneration, as well as the impotence of politicians, took part in founding deep hopes around computer programming. The utopia of the 'reactive environment' consequently proposed to rely on machines to better optimize and augment, calibrate and regulate the urban project in its systemic dimensions. MIT's heavy investment in behavioural sciences and computer science applied to architecture and urbanism strongly helped to create an urbanistic imaginary dimension of a group of tools oriented to projection, prediction and prescription based on data and computing.

The transposition of the simulation of information technology interactions to the urban was notably initiated by Jay Forrester, a specialist in defense systems, in his text *Urban Dynamics* in 1969. In 1971, when he was a member of the Club of Rome, he published *World Dynamics*, in which he developed a global modelling system that examined demographic growth and industrialization in a world with finite resources. The oil and environmental crisis would offer a new field of experimentation to 'system dynamics'. The Club of Rome commissioned MIT's Systems Dynamics Laboratory, headed by the very same Jay Forrester, a simulation of humanity's future over a period of more than a century. The report, *The Limits of Growth*, was published in 1972. In its predictions, nine scenarios led to going beyond the planet's bio-capacities and to collapse, and four others offered a more open future, with growth that would slowly stop. William Behrens, Donnella Meadows, Dennis Meadows and Jorgen Randers militated for the combination of several dynamics, the only ones that could reverse this collapse.[8] In 1992, in the update of the study titled *Beyond the Limits*, the authors formulated a

6 Halpern, 'L'architecture comme machine', p 130.

7 Nicholas Negroponte, *The Architecture Machine*, Cambridge, MA, MIT Press, 1970; *The Architecture Machine: Toward a More Human Environment*, Cambridge, MA, MIT Press, 1973.

8 William Behrens, Donnella Meadows, Dennis Meadows and Jorgen Randers, *The Limits to Growth*, New York, Universe Books, 1972.

new warning, this time with the observation that the limits of the planet's load capacity were exceeded. This example bears witness to a specific use of the tool: computer simulation was mobilized for a societal project slowing growth to the extent that it demonstrated that perpetual growth (notably economic) was incompatible with the preservation of the ecological system and human life, whatever the level of technical progress and innovation hoped for. This worldwide call to not *go beyond the limits* still powerfully resonates today. Paradoxically, however, the field of technological innovation will ceaselessly push back the limits of growth by using these same tools. The project of controlling stock and distributing information was replaced by that of capitalizing and storing data to accelerate economic flows. A new capitalization and extractivism principle was structured based on 'the data' (public and/or private) that appears as a new raw material. The economist Yann Molier Boutang, who developed the idea of cognitive capitalism, asserts in a succinct phrase that 'we are all Google's workers'. He adds that 'its principal source of value is intangible work, the capture of positive externalities, the condition of innovation and collective intelligence, the new resource. The muscular force specific to industrial capitalism would supposed have consequently made way for a new force, collective invention founded on knowledge-sharing.'[9] Let us recall that the largest revenue of the Alphabet/Google group is advertising: $219 billion in 2021, out of a total of $257 billion for the entire group. The data collected to be used, exchanged, shared and made profitable must be stored.

In the smart city product of the 2000s, the compatibility between a high-tech approach and the project of a more sober society was very little debated, and the environmental and energy impacts were not seriously evaluated.[10] The technical project of the smart city became that of a new age of maintenance, as François Ménard and Jean Daniélou pointed out in their study 'the art of augmented cities' in 2013.[11] The effects of the digital on the electricity sector demonstrate this with a certain clarity. With digitalization, the networks would become smart grids. From a prospective viewpoint, it is the promise of a perfect balance between production, distribution and consumption through bi-directional management in real time (which evens

[9] Yann Moulier-Boutang, *L'Abeille et l'Économiste*, Paris, Carnets Nord, 2010.

[10] The impacts of reduction are still rather negligible on the large scale. Deloitte Développement Durable, Eco Info, Futuribles and the CRÉDOC, *Potentiel de contribution du numérique à la réduction des impacts environnementaux. État des lieux et enjeux pour la prospective*, ADEME report, 2016.

[11] Jean Daniélou with François Ménard, *L'art d'augmenter les villes (pour) une enquête sur la ville intelligente*, La Défense, Plan Urbanisme construction architecture, 2013, Available from: http://www.urbanisme-puca.gouv.fr/l-art-d-augmenter-les-villes-pour-une-enquete-sur-a342.html [Accessed 12 November 2020].

out the points between producers and consumers). But many pilot projects are still in the test phase and it would seem that reality shows the difficulties of the electricity grid operators in following the call for consumption of the digital sector, notably data centres, rather than benefits of a more fluid and economical functioning of the grid itself through the effects of its digitalization. These effects are not collateral but central, if we refer to the difficulties of the sector's actors, namely the electricity transmission and distribution operators. It is when we return to the (infra)structures that we reach the limits of the virtual.

Criticism of the political and social implications of technical progress and the digital increased at the start of the 2010s. The researchers Evgeny Morozov and Francesca Bria were among those who called for going beyond the discursive coalitions of the smart city to abandon the production of homogenization and standardization processes and the consolidation of power logics, and to redeploy these tools to open a future capable of changing the Anthropocene paradigm and enrich metropolitan governance.[12] But what about infrastructures? The digital utopia continues to try to find its way notably through genealogical historic approaches like that of Félix Tréguer[13] in *L'Utopie déchue, une contre-histoire d'internet, XVe–XXIe siècle* or that of Fred Turner in his work *From Counterculture to Cyberculture:* Stewart Brand, *the* Whole Earth *Network, and the Rise of Digital Utopianism* that restores the connections of cyberculture at the crossroads of the military-industrial world, cybernetics and the counterculture. By using Steward Brand and his *Whole Earth Catalogue* as a paradigmatic figure, Fred Turner restores the dream of a horizontal and libertarian Internet as a metaphor of the territorial utopia of the American 'communalists'. It is one of the theses of his work: cyberspace would be the territory of utopian communities that were created, outside the material constraints of the real. Cyberspace is the place that links geographically distant groups: computers connected in a network enable individuals to collaborate in real time and to circulate information and knowledge. But this parallel curiously underestimates the material dimension of the digital, while the form and scale of the infrastructures (energy or digital) are a central subject of the debates of the counterculture movement, notably that of energy autonomy,[14] which places the networks' scale and structure in first place in the criticism of technologies. Fred Turner

[12] Evgeny Morozov and Francesca Bria, Rethinking the Smart City, Democratizing Urban Technology, New York, Rosa Luxemburg Stiftung, 2018.

[13] Félix Tréguer, *L'Utopie déchue, une contre-histoire d'internet, XVe–XXIe siècle*, Paris, Fayard, 2019.

[14] Fanny Lopez, *Le rêve d'une déconnexion, de la maison autonome à la cité auto-énergétique*, Paris, La Villette, 2014, translation *Disconnection Dream, from the Autonomous House to Self-Sufficient Territories*, Manchester University Press, 2021.

consequently sets aside the most technical-critical area of the American counterculture and confused the issue on the non-neutrality of technologies as though what was happening on the network was disconnected from its material structuring, whereas the digital technical system emerges as one of the most centralized ensembles.[15] It is difficult for us to find in the existing historiography the genealogy of an alternative digital system that treats user terminals, network infrastructure and storage as a decentralized ensemble that can be appropriated based on controlled interconnections. Of course, the first personal computers (PCs) and the suppleness of the networks connecting them can be distinguished from the more horizontal approach of the mainframes of cybernetics, but what about a system that, apart from computer technology that is a little 'garage punk', thinks of a much larger network? Paradoxically, this illusion of disconnection and this unhooking between user terminals (PCs and more largely connected objects) has constantly grown through a gradual centralization of storage and interconnection. The digital technical system has not escaped a strong monopolistic centralization and the illusion of creative destruction.[16]

---

[15] Jacques Ellul, Le Système technicien, Paris, Calmann-Lévy, 1977.

[16] Pierre Caye, *Critique de la destruction créatrice. Production et humanisme*, Paris, Les Belles Lettres, 2015.

2

The Internet Infrastructure: From Dispersion to Centralization

At the beginnings of the Internet in the late 1960s, digital infrastructures were rare and occupied very little space. Computers (mainframes), before being personal and portable, were enormous machines occupying whole rooms in a few universities, army research centres and a very few large companies. It was in 1973 that the first Arpanet network connected 38 computers in the entire United States. The territorial and spatial impact of this pilot project was very tenuous at the time. Starting in the 1970s, but especially in the 1980s, the personal computer heralded the end of mainframes and a deployment whose reticular spatiality was difficult to envisage at the time. In these beginnings, computer infrastructure was rolled out on a decentralized spatial principle. Before the Web, contents were hosted in each computer, on a peer-to-peer basis. It was especially with the development of the commercial Web in the late 1990s that the hosting of sites, messaging services and corporate platforms led to increasingly enormous storage needs. The Open Source and Free software movements, initiated by Richard Stallman in the United States and later by Linus Torvald in Finland, would remain defenders of this distributed scale, notably for data storage. The turning point (and the rhetoric) of the 'new economy' of dematerialization interested an increasing number of entrepreneurs and large companies. Centralization took place gradually and in a double movement: private digital actors emerged while the privatization of the public service of the telecoms was underway in the United States as in France. The development of optical fibre starting in the mid-1980s portended a network that gradually distanced itself from its connection with telecoms and copper, and confirmed the growing convergence between computer technology and telecommunications. The result of research largely conducted in the United States by the Corning Glass Works company, optical fibre permitted better and faster throughputs, better latency and avoided disturbances by electromagnetic waves. It was a network that emerged and was designed for the Internet. It induced a spatial hierarchy with the Internet backbones that comprised, in the 1980s, a long-distance, very high-speed link, starting from which the territories were served. However, this backbone idea sometimes refers to historic network infrastructures, but is no longer used because, since the 1990s, no network has officially been at the heart of the Internet. To make its way to the neighbourhoods and as far as the buildings, there were many combinations between an optical fibre backbone and local service using ADSL or VDSL, which were mostly copper networks.[1] In 1993, the Internet backbone of the National Science Foundation (NSF, an American public federal agency), the main backbone of the network, was sold to MCI, a private telecom company. Four sites then became network access points (NAP), in a transition from the public to the private sector: in Palo Alto, Mae East near Washington, DC, and in the cities of New York and Chicago.[2] These were the ancestors of the Internet Exchange Point. These NAP would then become territories of major digital concentration. They were the interconnection nodes of an immense web that was constantly expanding.

More centralized information systems developed in the service sector, offices, hangars, and places dedicated to hosting and computing. Growing obligations in terms of cybersecurity, linked to the development of spam and computer attacks, favoured the creation of firewalls and filters on the network.[3] The governments' determination to increase surveillance of the Web gradually brought about the growing centralization of the Internet. If reticular and infrastructural spatialization was still very heterogenous, the data storage question soon became unavoidable and this industry's market very quickly became structured. For Cécile Méadel and Francesca Musiani, it was when the Internet went from 'a tranquil utopia of passionate scientists to an agitated and powerful mass medium',[4] with the explosion of websites and new services, that the development of information technology architectures became centralized and required at the same time as increasingly powerful computing and data storage technologies. The 'data centre' sector specialized and became one of the sector's most strategic segments, the purpose of which was to host and process digital

1 It should be noted that optical fibre is not the only network that can offer high or very high-speed throughput (cable, VDSL2, DOCSIS), but it is the one that accumulates the most advantages.

2 These NAP are today still hubs for the Internet and the territories concerned: concentrations of data centres, Internet traffic and digital companies – each site having its own specialty.

3 Francesca Musiani and Valérie Schafer, 'Le modèle internet en question (1970–2010)', *Flux*, 3(85–6), 2011, pp 62–71.

4 Cécile Méadel and Francesca Musiani, *Abécédaire des architectures distribuées*, Paris, Presses des Mines, 2015, p 45.

data, but also to host network interconnection points (Point of Presence or Internet Exchange Point) and telecom operators. Data storage would initially be rolled out in existing buildings: former telephone exchanges, electronics factories or industrial buildings for the press, textiles, printing before becoming a very much standardized new building called 'telecom hotel', then data centre.

Digitalization was accompanied by infrastructural growth that can be divided into four segments: a large increase in user interfaces or connected objects; the development of network infrastructures themselves (terminals, cables and hardware, software); the deployment of specific infrastructures like data centres for storage and interconnections; and an augmentation of the traditional electricity infrastructure.

If the footprint of the digital technical system is equivalent to the appearance of major 20th-century infrastructures, it is more complex to grasp because its technical and architectural materiality is not as visible. The digital extends beyond classic infrastructural spatiality because its visibility is expressed in the virtual. The digital makes the gigantic boxes that host it invisible to the benefit of virtual worlds in which architecture is always more developed, complex and inventive: it is a network of links, pages and tabs that connect territories, creating a feeling of infinity and even, sometimes, exoticism. Starting off on a page on the perfect recipe for pizza dough, we are now able to visit the galaxies photographed by the Hubble satellite. Everything is without a stop, without changing platforms and especially with ever-richer interfaces, and will soon be in 3D if we believe the promises of the metaverse. Unlike road, railroad or even electricity networks, the digital gives us the illusion that we are travelling outside any axis. This is a paradox if ever there was one: the digital infrastructures have an enormous impact, a greater spatial and energy impact than an architectural infrastructure unlike the many other infrastructures that have entered the history of architecture. As the new 21st-century infrastructures, data centres are multiplying at great speed, but will they make beautiful ruins?[5] What architecture should be built, but especially to the glory of what or whom? The digital industry is largely private and maintains a fierce culture of secrecy surrounding its activities, but also discretion and anonymity for its ultra-secure infrastructures. Despite the clear desire for discretion and the fact that they can lodge themselves everywhere, data centres are the most visible spaces of the digital technical

[5] It is notable that it was Alexis de Tocqueville who made technical objects monuments to material culture. According to him, technical vestiges like aqueducts, reservoirs and electrical power plants gave a better account than any other object of the civilization that created them. Alexis de Tocqueville, *De la démocratie en Amérique II* (1840), chapter XII, Paris, Gallimard, 1961. On the technical monument, see also Dominique Rouillard, 'Le monument, la démocratie, l'objet', *Cahiers du CCI*, 3, 1985, pp 25–8.

system on the territory. In inserting themselves in the territory more or less far from the large companies that they make visible, in the end, doesn't the disclosure of data centres and their energy functioning destabilize the digital technical system and its associated economy that functions on the illusion of invisibility and distance?

3

An Infrastructure without Architecture?

Data centres are the most tangible spaces of the digital technical system on the territory, but remarkable architectures remain rare, far from an architectural tradition that is found in train stations, telephone exchanges and hydroelectric or nuclear power plants of specific formal and aesthetic characteristics.[1] At first glance, most data centres are boxes, hangars, sometimes camouflaged or decorated. The acceleration of flows has replaced storage architecture (brick warehouses and other stylized infrastructures) with more ordinary structures, boxes of all sorts, like containers[2] or huge hangars for storage of material or data.

The first Google data centre in The Dalles (Oregon) or that of Amazon Web Services, not far from Umatilla and Port Morrow (Figures 3.1 to 3.4) can be compared to this warehouse or factory typology.

The absence of architecture could be explained by three reasons. The first is that data centres are the almost organic extension of the server cabinets and computer rooms originally included in office buildings and university and research complexes. Storage was not thought of as an autonomous program, but as an outgrowth of the network. The change in scale happened gradually: a much larger cabinet, then a dedicated computer room. The initially decentralized character of the Internet favoured, moreover, many actors, and uncoordinated developments, without a global view and without any architectural consideration. The second was due to the rapidity of the development of the Internet, when the commercial Web was born in the 1990s, which favoured a certain opportunism in the choice of installing

1 Maurice Culot (ed), *Architectures de l'électricité. Architectures de l'âge industriel*, Paris, Norma, 1992.

2 Marc Levinson, *How the Shipping Container Made the World Smaller and the World Economy Bigger*, Princeton, NJ, Princeton University Press, 2008.

Figure 3.1: Entrance to the city of The Dalles

Source: © Fanny Lopez, 2018

Figure 3.2: Google data centre, The Dalles

Source: © Fanny Lopez, 2018

Figure 3.3: Entrance of the Google data centre, The Dalles

Source: © Fanny Lopez, 2018

Figure 3.4: Amazon data centre, Port Morrow

Source: © Fanny Lopez, 2018

server rooms and meet-me rooms (a kind of 'triage station', a place for the connection of cables, where the telecommunications companies physically connected to one another and exchanged data). The data centres needed for interconnection were hurriedly created near Internet Exchange Points (IXPs), strategic crossroads of the Web. In 1996, in Loudoun County in Virginia, the IXP, Metropolitan Area Exchange (MAE-East), brought together a very large quantity of telecom networks. To increase capacities, storage and interconnection spaces were then moved into 'a sort of enclosure in plasterboard, a space hollowed out in the building's underground garage, on the other side of the street'.[3] The geographer Bruno Moriset pointed out the emergence of a very specific real estate market at the start of the 1990s that played a fundamental role in the renovation and revitalization of a certain number of urban neighbourhoods, through the conversion of industrial sites, food or logistics warehouses in large cities like Chicago, New York and London.[4] Opened in 1990, Telehouse North was the first colocation data centre built in Europe. Installed in the Docklands district, it has been the historic hosting service provider of the London IXP since 1994. The third reason explaining the lack of architecture is that the sector's principal companies like Interxion, Equinix, Digital Realty and CoreSite (all created between 1998 and 2001) are real estate investment funds, primarily interested in a rapid return on investment of the facility. As a former executive of Goldman Sachs who worked at Digital Realty pointed out to us: 'The data centre is above all a real estate product with very high financial profitability.'[5] The profitability of the flow is based on the security of the installations (preventing loss, malevolent manipulation and data theft) and the good connection to the electricity and fibre networks.

Data centres present a few major characteristics in terms of form, volume, density and program. The imaginary dimension of the bunker (computer bunkers) persists. The data centre is an ultra-secure surveillance architecture, even if the typology of the campus makes it possible to sometimes push a part of the control systems to the exterior. Its envelope does not suggest the strategic importance of its content. We can, however, note a progression from the closed black box to a more controlled opening in its relationship to the exterior (windows on the façade, development of the surrounding area) and the interior (transparent partitions to see inside the data centre from

3 Andrew Blum, *Tubes: A Journey to the Center of the Internet*, New York, Harper Collins, 2012, p 66.

4 Bruno Moriset, 'Les forteresses de l'économie numérique. Des immeubles intelligents aux hôtels de télécommunications', *Géocarrefour*, 78(4), 2003, published online 21 August 2007.

5 Interview with David Schirmacher, President of 7×24 Exchange International, President of Reset Advisors, former Vice President of Operations for Digital Realty, New York, January 2018.

the hallways, etc). The space is generally divided into two parts: the office and logistics areas (with reception zones, offices, logistics and maintenance); and the computer and technical areas (the computer area is almost as large as the technical area, with undulators, cooling systems, backup generators, batteries, etc). It is an extremely cut-up space with few programmatic variations. Spaces are never seen that cross the core of the data centre in which the data cabinets (sometimes partitioned by the customer with its own security systems), circulations and technical spaces surround the central storage space, with office areas and occasionally a cafeteria.

The buildings constructed to hold data present several technical particularities: extreme robustness of the structures and floors, to support the load on the floor (weight of the servers and technical installations): between 500 kg and 1 T/m^2; adaptability for the routing of cables; great ceiling height (often 4 metres or more); flexibility of the interior spaces and large open spaces; and a minimum of posts. They are often single-storey rectangular buildings in areas where there is no real estate pressure, but they can frequently have four or five floors. Great height is rare except in certain American or Asian cities. Many of them are housed in existing buildings provided that they offer enormous spaces, even infrastructures: military sites, port authorities, former logistics zones, abandoned factories, which often already have backup generators, fuel tanks, water tanks, etc, are preferred by the operators. The separation of spaces and temperatures plays a key role in the infrastructure's efficiency. This concerns isolating climates and creating milieux: cool hallways, warm hallways, climate chambers, exhaust stacks, spaces for backup generators, various vats and tanks, cooling towers, undulators, battery rooms and command rooms. Data centres are extremely sensitive objects for the building's owner (in terms of security, accessibility and continuity of electricity service). The functional character is very standardized, which also explains a form of resistance to formal audacity. Energy or programmatic experimentations (like the hybridization of these infrastructures with other types of uses like offices, housing or cultural facilities) remain rare.

If the digital and cybernetics enriched the architectural and urban imagination of many groups in the 1960s, such as Archigram and its famous Computer City, no trace of storage spaces that were integrated into computer spaces at this period can be found. Starting in the 2000s, but especially in the 2010s, a period of new data centre construction, in dedicated buildings, opened. A few agencies developed great expertise on the question, like Gensler and CAC Architecture (in the United States), Enia Architectes, DK Architectes, Reid Brewin Architectes (in France), AGS Architects and Rosbach Architects (in the Netherlands) and NWA in Europe and Asia. More and more architects have become involved in this type of program, as can been seen in the exhibition 'Power House: The Architecture of Data

Centres', which was held in London at the Roca Gallery in February 2022. Among the projects presented, there was the resolutely high-tech building of Telehouse by the architect Nicholas Webb (NWA).

Data centres present a group of buildings with variable sizes and functions. They can be divided into two major types: those of companies and those of colocation. Company or administrative data centres host and manage their own data on their servers in a building that belongs to them. Colocation data centres (sometimes more rarely called infrastructure data centres) have more varied uses: hosting client company digital equipment (the operator supplies the space and electricity); availability of the hosting provider's servers and computer equipment for its customers (the customers can make temporary physical reservations for servers, disk racks, and network equipment in order to benefit from a guaranteed use that is not shared with the infrastructure); and reservations in Cloud mode (the customers can reserve virtual machines on the data centres' servers). There is, therefore, no (or very little) architecture, which does not prevent us from thinking of the mass-produced product and interior design. Creating architecture for a data centre or installing a heating network remains a constraint and an additional expense, only justifiable if the contracting authority for the installation makes an urgent request for it. Despite these general characteristics, the understanding of the typologies and the siting strategies have been structured these last ten years with a market that has been stabilizing.

4

Scales, Typologies and Siting Strategies of Data Centres

The results of a study carried out by the Lawrence Berkeley Lab[1] for the United States and that of ENER' France[2] made it possible to identify five size categories for data centres. The smallest are between 1 and 100 m^2 (often rooms in companies); the small data centres between 100 and 500 m^2, the mid-size data centres between 500 and 2,000 m^2; the large data centres between 2,000 and 5,000 m^2 and the hyperscales more than 10,000 m^2. None of these categories really takes into account the micro-data centres that are objects like Q-Rad radiators[3] by the French company Qarnot Computing and the Germany company Cloud and Heat, or Stimergy or d'Asperitas's digital boiler, which could be placed in the first category.

Mid- to large-size data centres have been heavily developing since the 2010s. For areas of 2,000 to 50,000 m^2, average electricity consumption varies between 1 and 25 MW (25 MW being the equivalent of the electricity consumption of a city of 50,000 inhabitants). We can observe a visible trend according to which these data centres are often larger today with an electricity consumption of 40–60 MW and the clusters they form sometimes make their consumption equivalent to that of the hyperscales. Hyperscale data centres have a very large area (> 10,000 m^2, > 10,000 servers) that corresponds to heavy energy consumption (> 25 MW, and often over 80 MW). They are often the only entities being hosted (a company), for example, a major Big Tech actor. Out of the 900 or so hyperscale data centres in 2023 worldwide, only a small number are installed in Europe,

1 Mohan Ganeshalingam, Arman Shehabi and Louis-Benoit Desroches, *Shining a Light on Small Data Centers in the U.S.*

2 The distinction varies: for the Lawrence Berkeley Lab, the small data centres are 1–100 m^2 and the intermediates between 100 and 2,000 m^2; for ENER', the small data centres are 1–500 m^2 and the mid-size data centres between 500 and 2000 m^2.

3 These are radiators that use distributed computer calculations.

mostly in Sweden, the Netherlands and Ireland. There were three in construction in France in 2024. It is important to mention the size of data centres because it is often asserted that it is the small data centres, which are numerous and not very efficient, that reduce the sector's global energy efficiency, whereas the energy performance supposedly increases with size (economy of scale and standardization), especially if the construction is recent. In fact, hyperscale data centres are new buildings with more rigorous environmental standards. On the other hand, the small data centres that are server rooms or cabinets dispersed in companies are extremely difficult to identify in France and generally use more energy. In 2003, the geographer Bruno Moriset stressed that 'the real estate infrastructures of the digital economy followed the plans of traditional location, with a preference for central neighborhoods in the outskirts of cities undergoing renovation'.[4] By traditional location, he meant 'business district'. Several phenomena have modified this geography since the early 2000s. The economy recovered from the bursting of the speculative dot-com bubble,[5] and the explosion of the digital and the development of the Cloud transformed every sector, with the emergence of industrial actors specialized in data production, processing and storage. The data centres covered a diversity of services and activities, which induced different location and development strategies. This type of data centre has a different spatial siting mode that will impact the territories according to particular modalities. The small data centres are distributed in urban territories and this trend will grow with the development of edge computing and the Internet of Things, which requires low latencies, as close as possible to users. Regional data centres, for small and medium-sized enterprises, local communities and universities are integrated into the urban network of mid-size and large cities, whereas colocation data centres are primarily located in major metropolises. Hyperscale data centres prefer to be sited in territories that are not dense, such as rural or peri-urban areas, in which land and electricity are abundant and inexpensive. For all the operators, two location criteria are genuinely structuring. The first is the proximity of the optical fibre Internet network, notably of the principal axes that are the Internet backbones, to connect to Internet Exchange Points (IXPs), platforms on which all the networks that distribute information in the world are connected to one another. The second is to have enough electrical power for its installation and possible development, in a stable

4 Bruno Moriset, 'Les forteresses de l'économie numérique. Des immeubles intelligents aux hôtels de télécommunications', *Géocarrefour*, 78(4), 2003, published online 21 August 2007.

5 After five years of continuous rise, the Internet bubble burst on 13 March 2000 with the precipitous fall of NASDAQ. The bursting of the bubble brought about the disappearance of a host of companies in the 'new economy'. This shock wave shook the major stock exchanges worldwide.

and reliable manner, and preferably inexpensive. This can be paired with the nearby presence of a substation.[6] These criteria are weighed against each other, according to the type of data centre operator. For example, land can be expensive if the stakes of being in a dense metropolitan zone, closer to users, is crucial. Thus, the Silicon Valley data centres, like those of Equinix, Interxion and Vantage, for example, are developed on onerous land, because proximity to the IT world and highly connected users brings them enough in return. Likewise, the French operator Zayo has a data centre in Vélizy, like in the centre of Paris (the Les Jeuneurs site). The sites are complementary: that of Paris is more constrained, but the prices there are triple those of Vélizy.[7] The price of land is high, but the return on investment justifies this, for the users, as proximity pays. As another example, a very large data centre can be installed in rural territories, like Amazon (AWS) in Umatilla (Oregon), along the Columbia River, or Facebook and Apple in the heart of Oregon (Prineville) have done, provided that they develop their private optical fibre network to connect to IXPs. Land is vast and affordable, but even if it is far from the users, fibre makes it possible to connect to them rapidly. The large data centres function in a complementary manner with those in metropolitan areas and take over once the connection is established with Internet users. However, we can observe that the hyperscale data centre actors, like Microsoft, AWS, Google and Facebook, have such large technical and financial resources that this can more easily support all these factors than colocation actors, data centres of proximity (called regional), large companies or universities. We must also cite a certain number of complementary location criteria: security, absence of nearby residential areas, reactivity of the hosting community in administrative matters, availability of renewable energies and specific rates for data centres, various tax incentives, abundant and affordable land, with the fewest constraints and easements possible.

This first part analyzed, from a spatial viewpoint, the social-technical reconfigurations generated by the mutations of telecommunications, from the telephone to the Internet. The digital takes part in making urban and territorial service networks evolve as an increasingly complex and multi-scalar assemblage driven by private actors whose intervention frameworks evolve in a hybridization and 'horizontalization' logic in a structural continuity

6 This criterion, unmeasurable until this point, could change in the future. Microsoft's unplanned experience in Clondalkin, just outside Dublin, shows that GAFAM can also take on their own energy production. Faced with the transmission network's incapacity to integrate the enormous rise in load of the hyperscale data centres, the company has been building, over three years, its own installation equipped with 16 gas generators for a total of 18 MW of power.

7 Interview with Frederick Coeuille, General Manager/Director Zayo France, Paris, 2018.

characteristic of the centralization of large-scale infrastructures. The appearance of data centres as an autonomous real estate program emerged at the start of the 1990s and its infrastructures have continued to grow. Whereas the architectural dimension remains weak, the territorial and energy impact of the phenomenon has ceaselessly been amplified.

PART II

A Globalized Territorial Continuum

The cornerstone of the digital technical system – the data centre – is not only a warehouse lost in rural zones, certainly very well-described by Rem Koolhas in his article 'Brave World: People, Places, Technology'.[1] This is only one part of the phenomenon, far from representing the complexity of the system that comprises a globalized territorial continuum, each piece of which is strongly intertwined with the others. If these storage buildings' strategy of visibility, form and programmatic evolution is trying to find its way, it is also the network dimension and the multiplication of these infrastructures that raise questions. They function in a complex interconnection system that links the small data centres to the large ones, those of colocation to enterprise data centre. The digital storage industry shows three major trends: growth, interdependence and ultra-centralization.

The first trend is the growth of data centres in every territory. The large American and Chinese Big Tech companies are continuing the development and construction of data centres on a very large scale, preferably in rural and semi-urban territories where land and energy are inexpensive. The second major trend is centralization. The analysis of the industrial mutations of globalization has largely showed relocations of flows in territories where labour and raw materials are less expensive, while keeping company headquarters in the large metropolises that comprise the network of world cities.[2] With the digital, relocation is limited and is concentrated on the national territories with dedicated hubs. The digital strengthens the hub effect.[3] Another phenomenon accentuating this concentration is the

---

1 Rem Koolhaas, 'Brave World: People, Places, Technology', *Flaunt*, 145, 2016.

2 Pierre Veltz, *Mondialisation, villes et territoires: L'économie d'archipel*, Paris, Presses Universitaires de France, 1996.

3 Dominique Rouillard and Alain Guiheux, *Le projet Hub – L'architecture des nouvelles mobilités*, Geneva, Métis Presses, 2020.

consolidation of the market's actors. In late 2019, Microsoft had 115 partner colocation data centres around the world and Amazon 125. In 2020, at a two-week interval, both of them announced that at the end of 2022, they would have no more than 20 partners worldwide and we can easily guess that in a few more years, this number will be around five.[4] The ranks are thinning. Consolidation means that there are fewer operating proprietors, but also in this precise case fewer buildings. This means that the obsolescence of certain data centres is already no longer a taboo, whereas the cable and networks will have been reinforced to supply them. The third trend, linked to centralization, is the growing interdependence of data centres. These digital configurations give rise to complementary spatial installations between rural, semi-urban and urban territories. The rural hyperscale function is in direct relationship with the large regional data centres or the metropolitan hubs, themselves connected to urban micro-storage sites. We can observe, today, the major development of gigantic data centres, in particular in the United States, China and Europe, which echoes the growing (and sometimes disparaged) concentration of the data of billions of people (3 billion, for example, for Facebook in 2023) by GAFAM and BATX[5] in China. These data centres are not located where the populations are concentrated, but where land and energy are cheap. The digital geography shows an urban concentration of flows via colocation data centres that are major interconnection hubs and entrances to Clouds and a geographic decentralization of stock with the hyperscale data centres. One of the effects is that the storage zones with a very low population density are becoming very dense in terms of electricity consumption, which strongly impacts their future development strategy.

This apparent contradiction between concentration and dispersion seems to mask several elements. The Internet network and the information flows have in fact followed urban hierarchies,[6] and favour metropolitan hubs. The spatial evolution of digital infrastructures demonstrates a growing process of concentrating data centres, and densifying the network, in a context initially of privatization, then of creation of large digital groups with a monopolistic

4 Interview with Fabrice Cocquio, President of Interxion France, Paris, September 2021.

5 BATX: Baidu, Alibaba, Tencent, Xiaomi.

6 Right from the beginning, the network was sited on strategic points in the city. It followed the historic networks. The Internet network first used that of the telephone, then optical fibre was inserted in existing ducts, along sewers, roads and highways (France), metros (Paris), railroad lines; or overhead with the electricity grid (like the high-voltage grid of the Bonneville Power Administration in the Northwestern United States or RTE in certain French rural territories). Consequently, following a path already laid out, the Internet strengthened urban hierarchies already in place, the cables connecting San Francisco, New York, Paris, London, Amsterdam and Frankfurt, economic capitals and metropolitan centralities of worldwide influence.

tendency. The development of the commercial Web and the privatization of telecoms in France and the United States thrust the digital infrastructures into a new era in which territories placed themselves at their service, with more or less public regulation. The spatiality of these infrastructures was therefore not planned and coordinated by the public authorities, neither at startup, nor in its strong growth phase. This almost entire absence of regulation and the predominance of commercial actors in the functioning of the digital infrastructures raises questions today of a political nature, on surveillance and control, and access to knowledge, as well as the populations' means of communications, but also on the spatial and energy impact on territories of every kind.

The development of data centres has gradually consolidated specific digital territories, on the one hand through their concentration on the major Internet Exchange Points (IXPs), increasing metropolitan polarization, and, on the other, through the dispersion of storage in the historic territories that serve the metropolises. In fact, the cutting-edge information technology needed for the Internet of Things requires micro-data centres as close as possible to users. It is also the small surface areas of the telecom operators, disclosed and described by Soline Nivet in her research for the operator Free[7] in Paris (on the ground floor or located in more imposing buildings), that shows this digital densification. The metropolitan zones are heavily affected by this phenomenon, and despite the move of certain colocation data centres outside the metropolitan areas, they remain very present. Their integration into the heart of the dense city raises new infrastructural questions each time this occurs.

In the United States, the largest concentration is unquestionably in northern Virginia, due to its proximity to the US capital, Washington, DC, and the historic presence of one of the first IXPs, Metropolitan Area Ethernet (MAE-East). Closer to a territory that serves the capital than of the effervescence of a metropolis, Loudoun County has almost reached 2.3 million square metres of digital storage area. If the city of New York concentrates less data centre area than before, notably since Hurricane Sandy in 2012, it is a major global financial centre and one of the largest IXPs in the world. It therefore constitutes a strategic gateway on the network, completed by data centres in neighbouring New Jersey. Next comes Silicon Valley, notably, the city of Santa Clara, then second-rank digital metropolises like Los Angeles, Chicago, Dallas-Fort Worth and Atlanta. In parallel, GAFAM have located their data centres in the more rural states of Oregon, Oklahoma, Ohio and Utah. In Europe, the four principal IXPs (Frankfurt,

7 Soline Nivet, *Architecture et numérique: Infrastructures, Interfaces & Institutions d'un 'écosystème' parisien*, HDR, Université Paris-Est Sup, 15 December 2020.

London, Amsterdam and Paris) gather the largest concentrations of data centres and the highest IT gross domestic product. In 2020, Dublin joined the leading quartet, with each time a strong increase in energy consumption that participated in weakening the national electricity systems. Faced with this pressure on the grids, the cities of Amsterdam and Haarlemmermeer decided to suspend their installation, waiting until the urban services and electricity operators define a long-term national planning strategy.

5

In the Dense City

The metropolitan centres, in the United States and in Europe, and even more so world cities like Amsterdam, London, Paris and New York, present very high population and job densities, and an exceptional concentration of wealth. These territories are strategic hubs in which Internet cables are installed and connected to one another; global and centralized decision-making centres in which multiple informational contents (cultural, financial, communicational, commercial, etc) are consumed and distributed. The former telephone exchanges have often become Internet Exchange Points (IXPs), paired with larger or smaller areas dedicated to data storage.

The historic New York locations

Following the terrorist attacks of 11 September 2001, but also after an explosion of the urban steam network that altered the entire computer system of Goldman Sachs in Lower Manhattan, the financial data centres, under the pressure of the US government, started to relocate to nearby New Jersey and northern Virginia, where energy and space are less expensive. The New York Stock Exchange also moved and since 2010 has rolled out almost 40,000 m^2 in Mahwah, in northeastern New Jersey. This data centre also offers spaces for trading companies, which are premium locations in terms of latency (every nanosecond counts). Despite high energy as well as real estate prices, but also because of climate vulnerability as demonstrated by Hurricane Sandy in 2012, which caused the shutdown of several data centres, New York has remained a city where you have to be connected to hundreds of telecom networks to have a gateway to the Clouds.

In New York, several buildings in Lower Manhattan have continued, with the Internet, their communicational destinies. A worldwide hub of exchanges, the data centres have discreetly inserted themselves into urban life through the transformation of the former telephone exchanges, but also industrial buildings from the early 20th century. What they all have in common is proximity with the Holland Tunnel, where most of the Internet

Figure 5.1: 32 Avenue of Americas, New York

Source: © Cécile Diguet, 2018

cables arrive, notably those that are transatlantic. Starting in the late 19th century, the city of New York welcomed the head offices of large telegraph and telecom companies. In 1914, AT&T (telephone) and Western Union (telegraph) both moved into 32 Avenue of the Americas (Figure 5.1), in an edifice built to the glory of worldwide communications. In 1928, Western Union moved a few hundred metres away to 60 Hudson Street (Figure 5.2) and had its head office built by the same architect who designed AT&T's

Figure 5.2: 60 Hudson Street, New York

Source: © Cécile Diguet, 2018

building, in the Art Deco style. These two former telephone exchanges were connected by an underground conduit through which a great deal of Internet traffic is routed today (Figures 5.3 to 5.8). In the late 1990s, the two buildings switched to the digital and, when their former tenants left, underwent a radical transformation, getting rid of telephone equipment to install IXPs, data storage spaces and cutting-edge telecommunications infrastructures. The Intergate Manhattan tower, a former telephone exchange located at the foot of the Brooklyn Bridge, did likewise. The

Figure 5.3: Arrival of cables in 32 Avenue of the Americas

Source: © Cécile Diguet, 2018

building was erected in 1975 by the NY Telephone Company (which later became Bell Atlantic,[1] then Verizon) before being bought by the

[1] Created in 1896, the NY Telephone Company was a merger of local companies. It then became a subsidiary of the American Bell Telephone Company, which became AT&T in 1899. Bell Atlantic came from the dismantling of AT&T in 1984 and following a merger with GTE, which became Verizon in 1998.

Figure 5.4: A conduit of cables

Source: © Fanny Lopez, 2018

Sabey company[2] (for around $100 million), and was rebaptized Intergate Manhattan. The telecom operator Verizon still occupies three floors today. Sabey invested $300 million for the building's renovation and the partial transformation of the data centre tower. It is 32 storeys high (160 metres),

[2] Sabey is a Seattle-based real estate company that has built a great deal for Boeing (warehouses, factories and hangars). It also specialized in 'critical facilities' for the military, but also for the healthcare and digital sectors with data centres.

Figure 5.5: The circulation of flows under the roadbed

Source: © Cécile Diguet 2018

Figure 5.6: 32 Avenue of the Americas, data cabinets

Source: © Fanny Lopez, 2018

Figure 5.7: 32 Avenue of the Americas, detail of a bay

Source: © Fanny Lopez, 2018

and offers 102,000 m^2 of available space. The ceiling heights vary between 4 and 7 metres according to the storeys and each open space proposes 3,600 m^2.

The transformation of these buildings initially erected for and dedicated to communications is very specific to New York. Construction particularities facilitated their transformation. The structure of the flooring supports the heavy loads of the servers; large open spaces offer areas for the deployment of bays and other generous technical spaces; their great height also facilitates

Figure 5.8: 32 Avenue of the Americas, safety cage

Source: © Fanny Lopez, 2018

the functioning of telecommunication antennas. Apart from telecom buildings, other former industrial, military, service and residential edifices were transformed to host data centres; this is the case of the emblematic 111 8th avenue in Chelsea (Figure 5.11). The fourth largest building in New York in terms of area, this building that takes up an entire block of nearly 275,000 m^2 is frequented every day by some 8,000 people. The former headquarters of the Port Authority, it was built in 1932 and bought in 2010 by Google for $1.9 billion, with its tenants including many telecom

Figure 5.9: Antenna and crane on the roof of 32 Avenue of the Americas

Source: © Fanny Lopez, 2018

operators and data centres. After having failed to evict its tenants, Google is still sharing the premises. Stanislas Voronov, manager of the Equinix site, recounts how, in 2010, faced with pressure from Google to recuperate the maximum amount of storage and interconnection space, a rumour was circulating that Google was going to become a telecom operator.[3] A few

[3] Interview with Stanislas Voronov, Manager of the Equinix site, New York, January 2018.

Figure 5.10: Sabey tower, New York

Source: © Fanny Lopez, 2018

years later, Google finally rented additional offices right across the street, at 85 10th Avenue.

Three subjects are particularly interesting: first, the recycling of existing buildings; second, the infrastructural compactness of these machine-buildings that contain all the equipment to ensure their autonomy (fuel and water reserves, cogeneration, heat storage); and, third, mixed uses – in each building, cited company headquarters, offices, radio studios, data centres and telecom operators cohabit. In Paris, the former telephone exchanges of the

Figure 5.11: 111 8th Avenue in Chelsea

Source: © Fanny Lopez, 2018

historic Postes, Télégraphes et Télephones (PTT), initially called Postes et Télégraphes (P&T), did not have the same destiny as the New York buildings. With the increase in Internet traffic and electricity consumption, most of the sites were no longer adapted and the transformation costs to integrate the backup electricity systems and air-conditioning were colossal.

At the heart of the Haussmannian fabric in Paris

In Paris, the telecom-built heritage of the beginning of the 20th century has mostly disappeared.[4] However, despite the policy of grouping together data centres begun by Orange and other operators, a few historic locations within the city itself still exist, notably in the former telephone exchanges of Beaujon (Paris 8th arrondisement), as the foundation of a residential building, and Montsouris (Paris 14th arrondisement) (Figure 5.12), which are partially dedicated to data storage. The principal IXPs are not located in these historic sites. These hubs are installed in the facilities of the colocation leaders, at Interxion, Equinix and Telehouse, and are located in Paris (11th arrondisement; Voltaire), Aubervilliers, Courbevoie and Pantin. The high

[4] Fanny Lopez, 'Les fantômes des télécoms: l'obsolescence des grands centraux publics', in Dominique Rouillard (ed), *Public – Infrastructure, architecture, territoire*, Paris, Beaux-Arts, 2021, pp 175–89.

Figure 5.12: Orange data centre in Montsouris in Paris

Source: © Fanny Lopez, 2019

connectivity of spaces historically linked to telephony located around the Opéra and the Paris Stock Exchange was perpetuated with the Internet, and the Sentier district, also called for a time Silicon Sentier, is the site of several large data centres that could move into adapted former industrial buildings. The data centres located in the heart of the Paris metropolis are therefore more transformations of formerly industrial buildings. The Telehouse operator moved, in the 1990s, into a building at 137 boulevard Voltaire, near the town hall of the 11th arrondissement, with a total area of 7,000 m^2 and available power of 5 MW (Figure 5.13).

This building, with its sturdy structure, housed until the 1980s a department store, Paris France, a competitor of Printemps and Galeries Lafayette. Nothing indicates today the presence of a data centre. However, it is a hub of the French Internet: it hosts the IXP France-IX; and is a central hub for 80 per cent of direct Internet traffic in France. One hypothesis for this siting is also the early presence of optical fibre in the area since the first link in France took place between the PTT telephone exchange at the Tuileries and that of Philippe-August in 1982. In the Sentier district, at 32 rue des Jeuneurs, Telehouse has 1,000 m^2 in the former Hôtel d'Agoult-Chalabre, a historic building with a metal structure erected in 1871 with a recent extension of 5 MW in a neighbouring building. The operator Zayo is very close, at 19–21 rue Poissonnière, on 1,250 m^2 (available power of 600 KW), also in a late-19th-century building with a metal structure, typical of the neighbourhood, developed for textile merchants and the press of the period.

Figure 5.13: Telehouse operator, boulevard Voltaire

Source: © Cécile Diguet, 2018

Free's data centre operator, Illiad, bought, in 2012, the building of the Ponts-et-Chaussées laboratory, on the boulevard Lefebvre in Paris (15th arrondissement), which includes a bomb shelter 27 metres below ground, which was built in 1949. At the time, the bunker was supposed to house the Ministry of Transportation in the event of an atomic bomb attack, in order to organize terrestrial communications. It now houses an 8,000 m^2 data centre. In Courbevoie, near La Défense, SFR installed a 12,000 m^2 data centre in the former factories of Delage automobiles, built in 1912 on 15 hectares. An enormous brick and metal building, topped with saw-tooth roofs, it would be the sole heritage marker at the end of the urban project underway in the Village Delage.

The industrial heritage has thus made it possible to ensure great connectivity and a digital hosting capacity in the heart of Paris and very close to its business districts (QCA and La Défense) in particular. Like Paris, Amsterdam favoured the urban development of data centres in dedicated zones.

The dedicated zones of Amsterdam

The Amsterdam metropolis has brought together very high-speed IXPs to which data centres have hastened. The Sciences Park Zone (in which the Amsterdam Internet Exchange, AMS IX, was founded in 1994) and the

Figure 5.14: AM4 (tower) and AM3 (horizontal building) of Equinix, Sciences Park

Source: © Fanny Lopez, 2020

Amstel III district (in the southeast), that of Schiphol Rijk (Haarlemmermeer commune), as well as the port of Amsterdam are particularly attractive sectors. Emblematic buildings were erected on the Sciences Park campus, which is a hyperconnectivity node, very close to central Amsterdam. It is a 70-hectare scientific research campus that houses teaching and research institutions. The campus has more than 175 companies and startups, most of which specialize in digital technologies. Between the buildings and university housing, the space for new constructions was limited and the price of land rather high. To obtain a good return on investment, it was therefore necessary for data centre operators to build a certain number of square metres vertically. There are currently three colocation data centres on the campus, one of Digital Realty and two of Equinix, which show the efficiency of the municipality of Amsterdam's efforts on spatial and architectural integration.

The AM3 and AM4 data centres of Equinix are surrounded by a water-filled ditch, which has replaced a fence (Figure 5.14). The Green Data Center award granted by Datacenter Dynamics distinguished the first building (built in 2012) in 2013. Connected by footbridges, the second building was inaugurated in 2017. With 4,200 m^2 of colocation space[5] and

5 AM4, Equinix, Available from: https://www.equinix.fr/locations/europe-colocation/netherlands-colocation/amsterdam-data-centers/ [Accessed 7 April 2021].

Figure 5.15: The data tower of Digital Realty, Sciences Park

Source: © Fanny Lopez, 2020

a gross floor area of 24,000 m^2, the buildings were designed by Benthem Crouwel Architects. Visible from the A10 highway, the larger building is 72 metres high. Divided into 12 storeys, it is sheathed by an openwork façade that accentuates verticality. Inaugurated in 2017, the Digital Realty data centre bears the symbolic name of Amsterdam Data Tower (Figure 5.15). Its 72 metres of height are divided into 11 storeys. It offers a floor area of about 16,000 m^2, of which 5,500 m^2 of space is for the customers.[6] Designed by the agencies AGS Architects and Rosbach Architects, its façade is covered with bronze-coloured perforated panels. To avoid fences, a glass façade was used on the ground floor to make the barrier effect seem invisible. The energy demand of 2 × 15 MW is ensured by a substation dedicated to the building.[7]

The construction of data centres in dense urban fabrics is rare today, priority often being given to housing or other programs. For the operators, it is simpler to develop in more adapted territories, with fewer urban, environmental and proximity constraints. They are found more frequently in metropolitan outskirts.

6 Datacenter digital realty, AGS Architects, Available from: https://www.agsarchitects.net/en/portfolio-item/amsterdam_data_tower/ [Accessed 5 March 2021].

7 Amsterdam Data Tower, Science Park 120, Amsterdam – Property Brochure, Available from: https://www.digitalrealty.com/ [Accessed 10 June 2020].

Figure 5.16: Façade of the data tower, Sciences Park Amsterdam

Source: © Fanny Lopez, 2020

6

In Urban Outskirts, Growing Digital Areas

Metropolitan outskirts, often industrial in the past, are ideal siting locations for colocation and Cloud data centres, and for local companies. It is a common development model: a formerly industrial territory, often serving the metropolis, offers large land takes and very substantial electrical power, but also good connectivity (optical fibre has followed highways and railroad lines, often crossing these territories). Data centres then replaced traditional factories. Is it the perpetuation of this service character or an upgrade? Each territory supports their arrival differently.

Stressing the single-function character of the economic territories on which they are installed, similar in terms of logistics or commercial zone landscapes, these digital activity zones are shifting towards urban fragmentation, with extremely large plots that cannot be crossed, often ultra-protected by defensive fences that do not contribute major urbanity to the territories. In France, in dense peri-urban zones like Saint-Denis and Aubervilliers, just north of Paris, and in the Netherlands in Schiphol and Haarlemmermeer (communes very close to Amsterdam), the development of data centres seems impossible to continue today for lack of space and available electricity. In the United States, the city of Santa Clara, in the heart of Silicon Valley, made them an economic development lever and a major and relatively integrated source of financing. In a different manner, the city of Hillsboro, near Portland (Oregon), preferred to take advantage of the relocation of its electronics industries to develop, in a sort of land opportunism, a digital zone otherwise called a data centre business park, rather similar to that of Dublin in Ireland; in both cases, enormous land reserves enabled a development dedicated to storage.

Santa Clara: a territory serving Silicon Valley

In the United States, Silicon Valley is the historic birthplace of the development of the Internet and information technology. It extends, in a dense and congested fabric, from San Francisco to San José with, at its centre, Palo Alto, headquarters of Stanford University, which houses a strategic and historic Internet Exchange Point for the West Coast. The Palo Alto Internet Exchange is located in a former telephone exchange (not the property of the operator Equinix) at the heart of the city, near its busiest retail thoroughfare, University Avenue. But it is the city of Santa Clara that has the highest concentration of data centres on its territory: around 60 that consume 70 per cent of the energy supplied by the municipal utility Silicon Valley Power (SVP), and represent 80 per cent of its revenue. Whereas the centre of the valley (Mountain View, Palo Alto, Cupertino, Sunnyvale) has offices and campuses linked to high-tech companies, in the south, Santa Clara and San José, in particular, concentrate on the storage infrastructures.

California was not the ideal destination for installations as strategic as data centres because of earthquake risks, high energy prices (other than in Santa Clara), and scarce and expensive land; but it had to be there. It was here that the headquarters of the largest actors of the digital world (Alphabet-Google, Facebook, Oracle, Apple, Salesforce) set up shop, as well as many startups developing new products, software and services and therefore collecting, creating, processing and storing data. It was also a dense metropolitan zone (for the United States), with a high-income population and developed digital practices – in short, a consumption and strategic prescribers hub. Silicon Valley consequently had a devouring need for data centres, and certain territories thus became its infrastructural back office.

Formerly a truck farming territory in the south of Silicon Valley, the city of Santa Clara industrialized throughout the 20th century, with the conservation of canning plants that processed agricultural productions, then, especially with the development of postwar military and computer technology research. Santa Clara then became a production city based on electronics and semiconductors with the head office of Intel.[1] The territory still offers critical electrical power, oversized networks especially for sanitation and large plots of land now sold at the price of gold, but above all the least expensive electricity in all of California. This advantage, undeniable for the data centre industry, pushed them to concentrate there, starting in the 1990s, transforming former electronics factories, then erecting their own buildings. Wendy Stone, Account Executive at Silicon Valley Power, 'loves' data centres. She states: 'They consume 70% of the energy supplied by SVP,

[1] Which has since moved to Hillsboro, in Oregon.

Figure 6.1: Data centres in Santa Clara; only Digital Realty is recognizable

Source: © Fanny Lopez and Cécile Diguet, 2017

representing 80% of its revenue and don't increase the territory's congestion since they create very few jobs.'[2] The city of Santa Clara has consequently benefitted from a strategic choice: that of having created its own electricity company in 1896, whose production units, notably hydraulic, have been amortized for a long time. It also supplies a dark fibre[3] network for companies, completing an ideal offering for the digital world. It is therefore paradoxically thanks to a public municipal electricity company that Silicon Valley's digital industries can function.

The valley has also seen the development of 'mute' service buildings, reflecting the culture of secrecy and local competition (Figure 6.1). Yen Chen, urbanist, Head of the Urbanism Department, city of Santa Clara, points out that 'before, everyone wanted to put forward its brand and its name, with an enormous logo and the colors of its graphic guidelines. But all that disappeared, even though the same companies are still there'.[4] Apple, Intel, HP and Microsoft all have laboratories and research and development centres and data centres in the city, but they are difficult to identify when walking around Santa Clara.

2 Interview with Wendy Stone, Account Executive at Silicon Valley Power, Santa Clara, October 2017.

3 Dark optical fibre is unused fibre that links two remote sites and enables the customer to transport large data flows.

4 Interview with Yen Chen, urbanist, Head of the Urbanism Department, city of Santa Clara, Santa Clara, 12 October 2017.

Figure 6.2: Vantage campus, Santa Clara; the structure of the new data centre being built

Source: © Fanny Lopez and Cécile Diguet, 2017

Standard architecture and no logo make the companies anonymous, thus minimizing the perception of the spatial and dominant economic presence of the sector. Historically in industrial zones, data centres have shifted to office zones (compatible with the neighboring habitat), which illustrates their mutation in terms of urban integration. Since 2015, however, land has become very scarce in Santa Clara and the space dedicated to data centres competes with housing needs.

Operators are now imagining developments that are more compact and with more architecture, like the Vantage campus that densified its plot of land by building higher on Walsh Boulevard (Figure 6.2). On this former Intel site, initially containing a single data centre, there are now six. The campus also has its own substation. The limit of urban regulations on heights is however 21 metres, although this can be negotiated on a case-by-case basis according the urban planner.[5] Faced with land pressure, the surrounding cities have taken up the challenge. San José, located just south of Santa Clara, is consequently hosting an increasing number of data centres: in particular, those of Coresite (Figure 6.3), but also Microsoft with a 40,000 m^2 building and new projects to come. Equinix also developed an enormous campus, on a former IBM microprocessor plant and farmland (Figure 6.4).

[5] Interview with Yen Chen, urbanist, Head of the Urbanism Department, city of Santa Clara, Santa Clara, 12 October 2017.

Figure 6.3: Coresite data centre, Santa Clara

Source: © Fanny Lopez and Cécile Diguet, 2017

Figure 6.4: Equinix data centre, San José

Source: © Fanny Lopez, 2017

Central Valley (east of San Francisco Bay) and the city of Hayward (where an Amazon data centre was built) are also developing data centres. And a little farther, a one-hour flight from Silicon Valley, the Hillsboro digital zone offers many advantages to the storage industry.

The Hillsboro digital zone

In the centre of the zone dedicated to data centres, the airport offers (in non-commercial flights) 32 private jets for the link with Silicon Valley. Near the city of Portland, Hillsboro is very dependent on the IT industry: in 2018, it concentrated 41 per cent of all digital jobs in Oregon. It also sees its salvation in the development of data centres, over 15 of which are concentrated in the north of its territory and is still developing on increasingly larger areas with Flexential and Digital Realty for example. About 140,000 m^2 have thus been delivered between 2018 and 2022. The French company OVH moreover moved there, but there is also Linkedin hosted by Stack, the Adobe software company and the Tata, QTS and Edge Connex data centres (Figure 6.5).

The oldest data centres moved into former electronics factories, the most recent ones are in new building but also display no logos. Despite the growing protest of the state of Oregon and Portland Metro (the only municipal organization of this type in the entire United States), due to the low rate of job creation and spatial sprawl, the city is slowly whittling away acres of truck farming land to host these infrastructures, spreading each time to the urban limit set by the metropolis.

Whereas the many industrial and port wastelands could house these digital uses in Portland, Hillsboro's connectivity on one hand and the tax exemptions on the other have consolidated this situation. '75% of the bandwidth of new transpacific cables under construction will arrive in Hillsboro in 2018',[6] pointed out at the time Mark Clemons, Director of Economic Development of the city at the time, strengthening a very strategic position on the Internet network nodes. Among its structural advantages, there is in Hillsboro as in all of Oregon abundant and inexpensive energy thanks to hydroelectric dams on Columbia River. These advantages included electricity infrastructures already well-calibrated for Intel, a large amount of available power and two new substations built by Portland General Electric, others by Bonneville Power Administration (Figure 6.6). The region's temperate climate favouring free cooling was also an advantage. Moreover, the cluster effect with Intel, the territory's historic actor that produces its processors there and the largest employer in all of Oregon (19,000 jobs) makes it possible to benefit from

[6] Interview with Mark Clemons, Director of Economic Development, Hillsboro, Oregon, October 2017.

Figure 6.5: The OVH and Adobe data centres in Hillsboro

Source: © Fanny Lopez and Cécile Diguet, 2017

Figure 6.6: Stack data centre in Hillsboro surrounded by power lines with, in the distance, the substation

Source: © Fanny Lopez, 2017

very advantageous tax exemptions: the city created an 'Entreprise Zone' that exonerates all the companies from property taxes. The municipality is also very welcoming for the business through its technical services, which facilitate all the steps concerning installation, development, contact and coordination with the electricity companies, the network, the state of Oregon and Portland Metro. Hillsboro has moreover created a working group dedicated to these one-stop-shop[7] questions for the data centres.

7 A one-stop-shop is an administrative or commercial location bringing together most of the services offered to the clientele.

Figure 6.7: Stack centre in Hillsboro

Source: © Fanny Lopez, 2017

Ted Reid, an urban planner at Metro, stresses however that 'tax incentives through special economic zones now requires data centres to create more jobs'.[8] Likewise, Paul Vaccaro, Vice President of Operations at the operator Stack, formerly Infomart, which has a large data centre, implies that this measure means that 'we don't want more of them in Hillsboro'.[9]

The local press relayed a certain number of questions faced with the granting of extremely advantageous taxation for them whereas they only create very few jobs, often a dozen or so for a 20 MW installation, without counting construction jobs. Stack for example benefitted from a 20-year tax exemption when it arrived. The tax effort to attract them is not equal to the economic benefits on the territories. Stack installed one of its largest buildings in a former electronics production plant built in 1998. It had a technical floor that Stack could reuse to install all the cooling machinery. An extension was added, which was created by the architect Jackson Main, the initial building's designer (Figure 6.7).

At the centre of the activity spaces, not far from the Intel factory and the highway stop, and alongside BPA's transformer station, this building offers 32,000 m^2 of information technology space. The electrical power of 18 MW

8 Interview with Ted Reid, Head of Regional Urbanism, Portland Metro, Portland, Oregon, September 2017.

9 Interview with Paul Vaccaro, Vice President of Operations at Stack, Hillsboro, October 2017.

is doubled by a backup reserve of the same power (generators). In 2018, Stack's main customer is Linkedin. The company occupies 30 per cent of the building and supervised all the installation of its servers and technical systems to attain a PUE of 1.06. The temperature is higher than in Stack's traditional colocation halls in the rest of the building: 22°C instead of 18°C. Constellation, Stack's electricity supplier, provides power from 100 per cent renewable energy sources (hydroelectric, wind, solar) through the Direct Access program that gives discounts to large consumers. Consequently, the cost of energy is 6.5 cents/KWh instead of 8 or 9 cents/KWh with a traditional operator like Portland General Electric (PGE).

The landscape of the activity zone on which the data centres are installed is maintained by the city and the owners and managers of the business park. It is generously planted, and the site has a park. The well-sized road network has many thoroughfares and is bordered by plantings and bicycle lanes. This American example can be compared to the Data Center Business Park in Dublin.

Grange Castle in Dublin, one of the largest clusters in Europe

At the end of 2018, the metropolitan region of Dublin represented 25 per cent of the global market share of the European data centre industry, its closest competitor, London, recording a market share of 24 per cent.[10] Since 2017, the data industry has increased by over 50 per cent. According to the Host report,[11] in 2017, there were 41 data centres in operation, with electrical power estimated at 420 MW. In 2019, the figure rose to 54, with a power capacity of 642 MW. Thirteen new data centres were completed in two years, with an average consumption of 17 MW each. In 2017, Host predicted that 1,000 MW of data centre capacity would be built by 2024. On the basis of the currently approved planning authorizations, this capacity should now reach 1,500 MW, and an additional 350 MW are envisaged. Their construction increased by 35 per cent. In 2020, ten data centres (the average size of each being 20 MW each) were under construction. This represents an investment of around €1.5 billion.[12] Tax incentive policies have made the capital's digital

10 Jones Lang LaSalle, EMEA 'Data Centre Q.4', *Market Review*, 2018.

11 Host and Bitpower, *Ireland's Data Hosting Industry 2019 Q3 Report*, Available from: https://static.wixstatic.com/ugd/914803_6e99282e6ce54b66ae04533408904c24.pdf [Accessed 28 February 2025].

12 The Dublin metropolitan area brings together the major actors of the Irish economy (53 per cent of the GDP). It is the country's principle economic growth driver (65 per cent of the growth of the GDP between 2000 and 2016); see OECD, Centre for entrepreneurship,

market a major segment of the country's economy. The South Dublin County Council's sector, about 15 km west of Dublin, is one of the largest clusters in Europe, bringing together all the digital industrial giants. In dedicated business parks like those at Grange Castle or in Tallaght built on former industrial or agricultural land takes, the enormous warehouses of Google, Amazon, Microsoft, Equinix, EdgeconneX, Interxion and soon CyrusOne are continuing their spectacular development, with sometimes problematic energy and environmental repercussions.[13] In 2019, Grange Castle Business Park had roughly 100 hectares of buildings dedicated to data storage.[14] Everywhere in the sector, cranes dot the landscape. Before, this local community, like most of the country, suffered from an economy slowed down by a lack of activities, which explains the relatively low price of land. The arrival of data centres in the county happened gradually after the siting of Microsoft's first hyperscales (Figures 6.8 and 6.9) and later Google (Figure 6.10).

The clustering effect is also explained by a proactive public policy to attract companies linked to foreign direct investments.[15] This evolution was possible through preparing the field and the regulations within the park by its owner, the South Dublin County Council (SDCC): prior zoning in the development plan was fundamental. Generous sizing ensures the availability of enormous spaces favourable to large land takes, as well as excellent electrical connectivity and dark fibre.[16] Before the massive arrival of data centres, in 1999, the US pharmaceutical company Pfizer had opened the way, by ensuring the sector's electricity supply through the construction of a first 23 MW substation. Since then, each large data centre has built its own. Historically, the owner of many activity parks in Ireland, the governmental Industrial Development Agency (IDA Ireland[17]) helps sell land, like, for example, that of Grange Castle, to investors. The SDCC plans to extend the 500-hectare park to the west and has announced the creation of new

SMEs, regions and cities, Available from: https://www.oecd.org/cfe/IRELAND-Regions-and-Cities-2018.pdf [Accessed 7 April 2020].

13 *The Irish Times* wrote in its article 'High-energy data centers not quite as clean and green as they seem' of 11 September 2019: 'The Irish state has largely attracted technological investments thanks to its tax incentive policies and by facilitating planning. However, we must pay greater attention to the legislative, planning and infrastructure conditions established by the state for these companies. … In parallel to the Planning and Development Act 2000 amendment that designated data centers as "critical infrastructures" permitting them to skip the local planning phase. The state has made every effort to support these companies to the detriment of environmental considerations. …'

14 Interview with South Dublin County Council team, February 2020.

15 Interview with Frank Nevin from the SDCC team, February 2020.

16 The dark fibre is an ultra-high-speed Internet solution for multi-site businesses.

17 Semi-public structure that has sought to attract very high-performance foreign direct investment industries since 1949.

Figure 6.8: Microsoft data centre, Grange Castle Business Park, 2020

Source: © Fanny Lopez, 2020

Figure 6.9: Grange Castle plot undergoing transformation in Dublin

Source: © Fanny Lopez, 2020

Figure 6.10: The Google data centre in Profile Park is one of the rare centres in the sector to distinguish itself with a mural on the façade, although this façade is practically invisible to the public (photo taken from private farmland)

Source: © Fanny Lopez, 2020

substations, two of which are currently planned in cooperation with the distribution and transmission grid operators ESB and EirGrid. However, the support for the expansion of the single-function data centre zone is no longer unanimous and Microsoft's expansion has resulted in long debates at the council. Our contacts in economic development, companies and tourism at the SDCC acknowledge that 100 hectares dedicated to data centres in the Grange Castle zone 'is a lot of land, a lot of energy for very few jobs, the balance of our community must also be ensured'. In 2016, the large electricity demand created destabilizations. Grange Castle found itself in an electricity shortage situation (four years earlier than anticipated) due to the massive consumption of the data centres. Natural gas was used while waiting for the network to be strengthened. Afterwards, the distribution and transmission operators strongly mobilized, notably, with the setting up in 2018 of the flexible demand.[18] It is a unique regulatory measure in Europe

[18] Interviews with Aisling Gilchrist, Engineer, EirGrid; Noëlle Ameijenda, Production Analyst, EirGrid; Tom Gallery, Data Centre Connection Manager, EirGrid; Shane Brennan, Specialist in Regulations and Contracts, EirGrid – all in February 2020. This flexible demand is particularly useful for data centres located in zones where electricity availability is limited. If a customer requests 80 MW, EirGrid does not provide the 80 MW directly, but provides what corresponds to real needs at the moment of the connection.

that makes it possible to offset the risks of over-investments linked to power reserved by the data centres. The firm capacity reserved for the data centre on the electricity grid corresponds to its real consumption after 18 months, and not to consumption hypotheses. Despite this type of regulatory tool, the continual rise in consumption in this sector is concerning. In a report written in July 2019, the Irish Academy of Engineering drew attention to the necessity of substantial investments in renewable energy production, as well as in the transmission and distribution grids.[19] In its conclusion, the report pointed out: 'The question of knowing who will pay for this investment and how it will be financed must be urgently discussed, likewise for the future development of the data industry itself, given the investment needs projected will be nearly 9 billion euros by 2027 to supply the electrical power needed for its development.' The government's declaration on the 'Role of data centres in Ireland's business strategy' recognized this problem, but until the present day, no measure has been implemented to guarantee that these costs will be entirely taken in charge by the data centre developers.[20] In September 2019, the Internet Ecosystem Innovation Committee (IEIC) drew attention once again to the risks of the strong growth expected and the impact this could have not only on the data centres' energy supply, but also on the local communities' economy. The IEIC called for a stronger and more restrictive constraint and planning policy for the digital industry, citing the Amsterdam moratorium as an example.[21]

Dutch saturation, the Haarlemmermeer and Amsterdam moratorium

In 2020, 75 per cent of the data centres in the Netherlands were located in the region around Amsterdam.[22] Nearly 2 million square metres of floor area should be dedicated to digital storage in the upcoming years, which would

The request is spread out and re-evaluated according to the increase in consumption. After 18 months, according to the consumption, EirGrid blocks a firm capacity. Consequently, the customer must precisely anticipate its Maximum Import Capacity, which is the maximum energy quantity it will use. If only 40 MW are used by the customer, the firm capacity is set at 40 MW: the remaining 40 MW are not reserved.

19 Irish Academy of Engineering, 'Electricity Sector Investment for Data Centers in Ireland', note, July 2019.

20 Irish Academy of Engineering, 2019.

21 Irish Academy of Engineering, 2019.

22 Interview with Petra Boegheim, urbanist and Head of Data Centre Strategy, Amsterdam, March 2020; Ruben Voerman, urbanist, regional and sustainability development, Head of Municipal Electricity Strategy, Amsterdam, March 2020; Tim Ruis, urbanist, regional development, sustainability, Head of Municipal Urban Heating Strategy, Amsterdam, March 2020.

Figure 6.11: Data centre in the Schiphol Rijk zone

Source: © Fanny Lopez, 2020

almost double the current capacity according to the Dutch Data Center Association.[23] In 2018, the metropolitan region housed 47 colocation data centres and 53 installations with a single tenant,[24] the majority of which were located in Amsterdam, Almere, Aalsmeer, Haarlem, Hoofddorp/Schiphol (Haarlemmermeer municipality) and Purmerend (Figures 6.11 and 6.12). In January 2020, CyrusOne announced the opening of the company's first data centre campus in the Netherlands in Schiphol, with expected power of 72 MW. Among the reasons explaining the siting of foreign companies and data centres in the Netherlands are, primarily, tax advantages[25] and the location of the metropolis that serves as a gateway to continental Europe. Just southwest of Amsterdam, the city of Haarlemmermeer has the Amsterdam-Schiphol airport within its territory. In this high-connectivity zone, almost all the vacant logistics land has been monopolized by data centres. Historically, the city of Haarlemmermeer could force better urban integration on the data centres located in the Schiphol Area Development Company activity zones, of which the city owns 25 per cent of the land.[26]

---

[23] Interview with Stijn Grove, Director of the Dutch Data Center Association, March 2020.

[24] *North Amsterdam Data Center Campus, Economic Impact Study*, report, June 2018.

[25] The tax rate on corporate revenue in the Netherlands is 19 or 25 per cent (according to taxable profits).

[26] Interviews with Mariëtte Sedee, Burgomaster, regional development, the environment and agricultural affairs, city of Haarlemmermeer, March 2020; Jeroen Graafland, Head

Figure 6.12: Data centre in the Schiphol Rijk zone

Source: © Fanny Lopez, 2020

Starting in 2017, strong electricity tension was felt on the grid, forcing the city to react by a moratorium. On the initiative of Haarlemmermeer, followed by Amsterdam, the decision was made in 2019 to prohibit new constructions of data centres[27] on their territory for the duration of one year. During this building permit suspension period, the municipalities started reflections and work with the operators of the energy transmission and distribution grids. These collaborations gave rise to a roadmap to initiate a new energy and spatial planning project on different scales of the territory (local, regional and national). On the municipal scale, the new data centres would have to adhere to a reuse perspective. They would have to be built so that their structure could serve other uses and/or their building materials could be reused. They would also have to be integrated into the environment and the ground floor would have to be shared with other functions, such as restaurants or commercial premises.[28] Haarlemmermeer's policy was expressed by a delimitation of the zone on which the data centres were

of the Urban Planning Department, city of Haarlemmermeer, March 2020; Clément Torre, urbanist, city of Haarlemmermeer; Irma Graveland, city of Haarlemmermeer, all in March 2020.

27 Apart from those to whom the building permit had already been delivered.

28 City of Amsterdam website, Available from: https://www.amsterdam.nl/bestuur-organisatie/college/wethouder/marieke-doorninck/persberichten/amsterdam-selectieve-groei-datacenters/ [Accessed 15 January 2020].

authorized to be installed. The legal expression of this policy is found in the Umbrella Plan Datacenters (*ontwerpbestemmingsplan Parapluplan Datacenters*) zoning plan. This plan indicates a 'quota' of hectares that the municipality can provide for data storage, where and under what conditions. The data centre policy of the municipality of Haarlemmermeer was presented to the council at the same time as the zoning plan project in October 2020, and was eventually validated.

This urban peripheral zone shows a few similarities to other infrastructures on the outskirts of the cities. This is notably the case of Plaine Commune, right outside Paris, which has the largest concentration of French data centres. Wasteland in a sector now saturated by these storage infrastructures has been constantly developed over the last ten years.

Plaine Commune, the leading French hub

The Plaine Commune conurbation is located just north of Paris in the Seine-Saint-Denis department. Plaine Commune[29] is the French territory with the largest concentration of data centres, Saint-Denis and Aubervilliers being in the lead with respectively five and eight buildings. According to Alain Vaucelle, Head of the ICT Mission for Plaine Commune, the conurbation had about 20 data centres in 2018. By 2021, according to Fabrice Cocquio, President of Interxion France, there will be 47 data centers covering 25 km^2 of territory.[30] Their presence dates back to the 1990s. The first colocation data centre was in Aubervilliers, in the park at the Portes de Paris, and was opened by Interxion in 1999. The presence of enormous plots of land available at affordable prices in the immediate proximity of Paris and with good electricity availability, witnesses of the industrial past of the Plaine Saint-Denis favoured the development of these infrastructures at a time when development in Paris was becoming limited.

The concentration also resulted from the quality of connectivity, with the presence of one of the principal axes of the Internet network located along the A1 highway, enabling the link between different European data centres. This grouping was also the result of the optimization and pooling of dark

29 The territorial public establishment Plaine Commune is an intercommunal structure that brings together Aubervilliers, Épinay-sur-Seine, L'Île-Saint-Denis, La Courneuve, Pierrefitte-sur-Seine, Saint-Denis, Saint-Ouen-sur-Seine, Stains and Villetaneuse.

30 Interviews with Fabrice Cocquio, President of Interxion France, Paris, March 2018 and September 2021. It must be recalled that the data for enterprise data centres are difficult to access and sometimes vague: when certain will count as one the data centre sites that have several buildings, others will count each building as a separate data centre. Some can even talk about colocation data centres without specifying it, whereas there are also enterprise data centres.

fibre costs.[31] There are 16 well-identified data centre sites for seven operators, with, in the lead, Equinix and Interxion (which recently merged with the American giant, Digital Realty), European leaders of the colocation market, both with nine sites. Equinix also announced, in January 2022, a new data centre on an existing site in Saint-Denis, near the municipal swimming pool that it plans to heat, for a total investment of €135 million. In a relatively heterogenous urban environment, the buildings in Plaine Commune are large, their average area varying between 1,000 and 5,000 m^2. Among the most imposing buildings is that of Interxion in La Courneuve, on the rue Rateau, which, after extension, now reaches 9,000 m^2 of computer rooms, and the Equinix data centre[32] in Saint-Denis, with 13,000 m^2 of computer rooms on a plot of nearly 6 hectares.

A slowdown in the number of data centres being built has been observed since the 2010s, a phenomenon that can be explained by three principal factors. The first is that land is no longer as available and inexpensive as in the preceding decade, notably because Plaine Commune has pursued a substantial urban transformation and promotion dynamic, with notably the development of the Campus Condorcet, the arrival of line 12 of the metro at Front Populaire and residential developments in the northern part of the sector of the Entrepôts des Magasins Généraux de Paris. The second is that the colocation sector has consolidated and the different operators have bought companies and existing buildings over these last few years. Consequently, Equinix, the world leader in colocation, bought companies like Telecity, while Digital Realty Trust and Interxion merged in 2020. The third factor in the operators' anticipation. A large number of installations were oversized to reserve market shares. Some of the installations have therefore remained vacant and are being filled gradually. However, the sector's dynamism is continuing, as shown by the delivery in 2022 of Interxion's 40,000 m^2 building in La Courneuve, on the former Eurocopter site, announced as the largest data centre in France. In 2022, Plaine Commune thus has some 84,000 m^2 of interconnection and storage space and about 360 MW of electricity dedicated to the functioning of these infrastructures.

If the 2013 ALEC study had warned local elected officials about the data centre phenomenon and had made the subject a priority on the eve of the 2014 municipal elections, mobilization has somewhat slacked off since. This is notably due to the slowing down of these storage infrastructures'

[31] The terrestrial telecom networks, when they supply the buildings, spend on average €3,000–4,000 per linear metre. For a business case to function, several data centres must be amortized. The contiguity and proximity of data centres is not a brake on business development – on the contrary, it is an accelerator.

[32] PA 2 and PA 3 are in fact on the same site.

construction rhythm, but also because of the many large-scale projects in which Plaine Commune is involved (the 2024 Olympic Games, urban renewal, Grand Paris Express). Therefore, in the end, the question has not really been discussed on the intercommunal scale, whereas other subjects have working groups between the communes (like cyclable mobilities, for example). Nonetheless, the conurbation has not forgotten that data centres raise problems. First, the risk of shortages for the local electricity system. The electricity distribution operator Enedis has just delivered a new substation to Aubervilliers, near the Campus Condorcet, in order to better meet requests for connections, but these capacities are already totally reserved. Citizen opposition and mobilizations concerning the siting of the Interxion data centre on the rue Rateau in La Courneuve received some media coverage when it was being built. Noise pollution raised by the inhabitants, the dangers linked to fuel oil storage and electromagnetic waves raised difficult questions concerning the municipality,[33] whereas the reservation of one of the last large spatial land takes of the territory – the 6.3 hectares of the Eurocopter site – was in 'negotiation' with Interxion for a very low number of jobs (Figure 6.13). The large-scale project of Interxion in La Courneuve is an illustration of the difficult dialogue between actors[34] and conflicts of land use in a territory where the shortage of housing, among others, is an important issue. In 2013, Interxion filed a first building permit for the construction of a storage building on only one part of the site (4.3 hectares). Carried out by the Anyoji et Beltrando agency, in partnership with the commune and the conurbation, the study projected developing two spaces:[35] one dedicated to the Interxion data centre along the highway, and, in the northern part of the plot, an urban complex with housing, to complement he Chabrol sector, its existing facilities (football, dojo stadium) and other new housing developments (Figures 6.14 and 6.15). The local urban plan

[33] See the chapter of the thesis by Clément Marquet on the opposition battle in Plaine Commune: Clément Marquet, *Binaire béton: Quand les infrastructures numériques aménagent la ville*, under the direction of Jérôme Denis, thesis defended on 30 October 2019 at Université Paris-Saclay (ComUE), in the framework of the École doctorale Sciences de l'Homme et de la société in partnership with Télécom Paris (Palaiseau) (thesis preparation establishment) and the Institut interdisciplinaire de l'innovation (Paris) (laboratory) .

[34] At the announcement of the departure of Airbus-Eurocopter from the factory, located at 13 avenue Marcel Cachin, the site was immediately bought by a local investor, without the community exercising its right of urban pre-emption or having the plot transferred to the Établissement Public Foncier d'Île-de-France (EPFIF) in order to be able to control its evolution. The site is located near the city centre, opposite public facilities (for example, sports, high school) used by many inhabitants, in a context that is currently hostile to pedestrians.

[35] Agence Anyoji et Beltrando, *Cahier de prescriptions urbaines, paysagères et architecturales, étude de programmation urbaine et économique*, December 2012. Archive du service de la ville.

Figure 6.13a: Site buildings

Source: © Fanny Lopez, 2018

Figure 6.13b: Site buildings

Source: © Fanny Lopez, 2018

Figure 6.14: Plan of the site with the data centre plot

Source: © Anyoji-Beltrando architectes et Urbanistes, 2018

of La Courneuve approved this project and zoned the site according to the uses that were decided.

In 2018, Interxion filed a new building permit revoking a certain number of decisions made with the local communities, since in this latest project the operator planned to develop a data centre on the entire 6.3-hectare plot, one of the last large land reserves in Plaine Commune, according to Fabrice Cocquio.[36] Apart from its size and connectivity (fibre along the A86 highway), the presence of RTE power lines that are routed along the plot (along the railroad tracks) is strategic. It was therefore possible to directly

[36] Interviews with Fabrice Cocquio, President of Interxion France, Paris, March 2018 and September 2021.

Figure 6.15: View of the project, 2018

Source: © Anyoji-Beltrando architectes et Urbanistes, 2018

connect to the RTE grid and for the data centre to free itself from the Enedis grid, the power requested being 130 MW[37] for 40,000 m² for information technology and an investment of over €1 billion. Interxion also invested in an 80 MW substation. The conurbation was then tempted to start new discussions to improve the integration of the data centre (fences, entrances, plants). Interxion finally proposed a public garden of 7,500 m². The circular building over 20 metres high was partially delivered in the summer of 2021 (Figure 6.16).

The digital activity zones have spatial effects rather similar to those of logistics zones. They show a hybridization between industrial functions on the one hand (massive electricity consumption, fuel oil reserves, etc) and service functions with the presence of offices and meeting spaces on the other. They are, therefore, rather well-integrated into spaces like the Plaine Saint-Denis, in which there is a heterogenous fabric comprised of former industries, and service and residential premises, as well as dedicated service spaces like those in Santa Clara, Hillsboro and Dublin, in which the business park form dominates. The accumulation of data centres and the consolidation of these 'digital activity zones' sometimes raises questions, to the extent that they supply a model (the activity zone) that seems obsolete with regard to

[37] Interview with Fabrice Cocquio, President of Interxion France, Paris, March 2018.

Figure 6.16a: New Interxion data centre in La Courneuve

Source: © Fanny Lopez and Maximilian Gawlik, 2020

Figure 6.16b: New Interxion data centre in La Courneuve

Source: © Fanny Lopez and Maximilian Gawlik, 2020

the urgent requirement needed to fight against the effects of climate change, but also because they create strong pressure on the local energy grid. What these spaces have in common is to propose single-function territories that do not favour complementarities between uses, but sometimes by minimizing conflicts. With large, very mineralized land takes that cannot be crossed and that are rather hostile to pedestrians, these dedicated buildings seem difficult to integrate, whereas their proximity to the dense city could permit energy pooling as we will see in the later chapters in this book. The scale of the infrastructure raises a question here, in the same way as the conflicts between operators and local communities. With any strong public regulation, these difficulties are also found in rural zones, despite the apparent ease due to land and electricity opportunities.

7

Isolated in Rural Zones

The operators of large data centres are very present in the rural world whose communities looking for an economic boost offer isolation, land availability and advantageous taxation. In terms of development, the sprawl principle dominates, in a sometimes imbalanced relationship between operator and local community, which is notably the case in Prineville, Oregon. In Europe, the installation of very large data centres is more complex, but the lack of space is compensated for by the very strong regulation of the public authorities, as we will see in Sweden, in the forested territories of Gävle and Sandviken and in the Netherlands, north of Amsterdam, in the agricultural zone Hollands Kroon.

Facebook and Apple in Prineville, Oregon

The dispersion of infrastructure in the immensity of large American territories has been a fruitful strategy for Big Tech, which can install its own fibre to connect to Internet Exchange Points. However, there are few companies and even fewer data storage operators able to pay for the luxury of discretion, paired with almost infinite spaces. 'When Facebook started to look for a site for its factory, their prospector presented the project to us under the name of Vatas. We had no idea what Vatas was, except that they wanted 80 to 120 hectares, that they were going to use a lot of electricity and water.' Betty Rope, mayor of Prineville,[1] a peaceful town in the heat of Oregon's high desert, added that the exact same thing happened two years later with Apple, installed right next to the city's heights, invisible from afar: first the mystery and the code name, then the obligation for confidentiality until its inauguration. More carefully thought-out than Google's first data centre sited along the Columbia River in The Dalles, the Apple (Figure 7.1) and Facebook (Figure 7.2) buildings are the reflection of their obsession with

[1] Interview with Betty Rope, Mayor of Prineville, Prineville Oregon, September 2017.

Figure 7.1: Apple data centre in Prineville, Oregon

Source: © Fanny Lopez, 2017

Figure 7.2: Facebook data centre in Prineville

Source: © Fanny Lopez, 2017

security and opacity: roads closed by concrete blocks, double rows of fences, along with drainage ditches, video surveillance, and permanent patrols by security guards in pickups.

If the Facebook logo is discreetly visible, its flag no longer floats in the air, and the white Apple logo is nowhere to be found. Half-covered by a gate covered with the juniper bushes that invade the region, the lines of the buildings almost blend with the horizon.

In 2021, Facebook announced that it wanted to reach a total of 11 data centres in the next years, or 430,000 m^2 in total or twice the size of Grand Central Station in New York or more than 17 times that of the Champ de Mars in Paris. The three façades of the first Facebook complex are as long as that of the Château de Versailles on the garden side. Paradoxically, their monumentality is not a show of power or a demonstration of force, but it especially the mirror of the excessiveness of contemporary digital practices. However, the mayor of Prineville considered in 2018 that Facebook 'was a good, a very good neighbor'. With Apple, they could redynamize an economy completely destroyed by the 2008 crisis (with an unemployment rate of 21 per cent), creating some 500 direct and indirect jobs, notably with construction sites. Facebook provided its employees with a great many services, and even proposed, according to Phil Steinbeck, 'the city's best restaurant'. As Jenna Burell states it, the Prineville data centre deal is 'situated in a much longer local struggle over rural development, recovery, and revival in the wake of dwindling natural resource industries'.[2] No one, however, could have imagined that the electrical power of the city would leap in a decade from 10 to 500 MW. Facebook and Apple consequently invested in five 15 MW solar farms, not far from their sites (Figure 7.3).

Each operator pays for the expenses of annual projects, in this contributing to investment in the local infrastructures (substations, water production, sewer plant). Lastly, Facebook, more 'community-oriented' than Apple, 'reserved and discreet', according to the mayor, willingly takes part in financing free summer concerts in the park, new IT tools in schools (iPad for everyone) and school scholarships, for a total, since its arrival, off over a million dollars. Nonetheless, as the live local Oregon media has stressed: 'The local tax relief permitted Facebook to save nearly 130 million dollars between 2012 and 2020, according to Crook County records.' Here, the digital giants appear as an opportunity, a job and economic development vector. Even the electricity companies court them. They are perfect customers: stable large consumption day and night, all year round (no peak to manage) and in constant growth since, according to Anish Gautom, engineer at

[2] Jenna Burrell, 'On half-built assemblages: waiting for a data center in Prineville, Oregon', *Engaging Science, Technology, and Society*, 6, 2020, pp 283–305.

Figure 7.3: Apple solar farm in Prineville

Source: © Fanny Lopez, 2017

the California Energy Commission, 'their consumption doubles every 18 months'.[3] As the energy efficiency programs start to bear their fruit, household consumptions drop and it is too costly to finance new electricity production and transmission infrastructures, the digital is a windful that is counted in gigawatts (and therefore in dollars) for the old utilities. According to Matthew Chancellor, Regional Manager of PacifiCorp, the historic company of northern California and southern Oregon, which supplies energy notably to Facebook's data centres in Prineville, 'the electricity companies love data centers!'[4]

A little farther north, from Port Morrow to Umatilla, once again near the hydroelectric manna of the Columbia River, Amazon Web Services (AWS) has rolled out storage warehouse that resemble factories, in gloomy industrial parks or rural no man's lands. Even more discreet, even truly opaque, Amazon is building data centres at a frenzied pace throughout the world. A former project manager, who has since moved to Zayo, the American telecom and data centre operator, said that he wore himself out on the four corners of the planet building identical buildings for the

[3] Interview with Anish Gautam, Engineer at the California State Energy Commission, September 2017.

[4] Interview with Matthew J. Chancellor, Regional Manager, PacifiCorp, Bend, Oregon, September 2017.

Figure 7.4: Amazon data centre in Umatilla

Source: © Fanny Lopez, 2018

e-commerce giant, now also the leader of the Cloud.[5] No care was taken with the building, which could be taken for a food storage warehouse or logistics platforms, topped by imposing air-conditioners and hooked up to powerful backup generators running on diesel oil. But there are no large trucks nearby or unloading docks, as the goods discreetly go through the optical fibre network. The installations still under construction in 2017 are heavily monitored by a local security company, and the name of Amazon cannot be found anywhere (Figure 7.4). Umatilla, a small lackluster town of almost 7,000 inhabitants, born at the same time as the gold rush in 1863, has housed, since 1962, stocks of chemical weapons for the United States Army. It has its own municipal electricity company and did not say no to Amazon when it wanted to build a site there. Julie Peacock, at the Public Services Commission of Oregon, points out that small towns 'want data centers to come, but this risks endangering their own electricity grid'.[6] The demand in power was so great that the company committed to building new production units for which, on one hand, it would go into debt and, on the other, passed along the cost to all the consumers of its

[5] Interview with Fred Jalali, Energy Manager of the Zayo data center in Santa Clara, California, September 2017.

[6] Julie Peacock, Principal Advisor on Policies, Oregon Public Utility Commission, Portland, Oregon, September 2017.

territory, even though they had modest incomes, and especially without any assurance that AWS would be there five years later. The company thus exerted pressure that could destabilize the economic and energy balance of the territory. Crystal Ball, Head of Public Relations Bonneville Power Administration (BPA), further stressed that while BPA tried to find alternatives to the heavy investments on its high-voltage power grid to improve its functioning, Big Tech used its influence to add transmission capacities and thus ensure the growth of their activity, to the detriment of the public interest.[7] However, despite the small number of jobs created, as well as the discounts negotiated on both state and city taxes, rural communities continue to want to welcome new data centres to improve their often dramatic economic situations.

Letha Tawney, Innovation Director for Renewable Energies, stresses that the American digital industry is the first to be collectively organized to make the most of the energy transition through organizations like the Renewable Energy Buyers Alliance.[8] If Amazon is one of the bad players criticized by Greenpeace in its 2017 Click Clean report, Facebook is on the front line in several states when it comes to negotiating advantageous rates for renewable energies in the end serving the entire digital storage industry. Paradoxically, in addition to the gains in energy efficiency that data centres acquire, the growing use of renewable energies can in this way counterbalance the excessive production of digital data. Consequently, the weakening of local electricity transmission and distribution infrastructures to the detriment of other needs of the populations and industries could over time threaten the endless expansion of the digital industries. In Europe, the presence of hyperscale data centres is lower in rural zones, and it is limited in a few countries. Their installation is however, strongly increasing.

Hollands Kroon in the Netherlands, greenhouses and hyperscales

With their landscape composed of enormous fields and greenhouses, canals, artificial islands and polders,[9] the Netherlands today is the archetype of a totally anthropized territory as far as its most rural boundaries. Google invested a billion euros in the Netherlands. The American giant built a

[7] Interview with Crystal Ball, Head of Public Relations Bonneville Power Administration (BPA), Portland, Oregon, September 2017.

[8] Interview with Letha Tawney, Innovation Director for Renewable Energies, September 2017.

[9] Land reclaimed from the sea, often located below sea level.

Figure 7.5: Part of the Microsoft data centre in Agriport A7

Source: © Fanny Lopez, 2020

data centre in the commune of Hollands Kroon in Agriport, north of Amsterdam, and enlarged its complex in Eemshaven, near Groningen. The two hyperscale hubs developed due to inexpensive land, an attractive taxation, energy availability and underwater cables. Bordered by the A7 highway, the Agriport A7 zone,[10] created in 2006, has an area of 1,000 hectares including enormous greenhouses, logistics warehouses, data centres and energy production infrastructures (wind turbines and gas cogeneration plants). The installation of the first Microsoft data centre (Figures 7.5 and 7.6) was made possible through the cooperation of Agriport A7, the municipality of Hollands Kroon, the North Holland province, the development agencies of the country and the northern region of Amsterdam.[11] It was in 2013 that Microsoft announced its project of investing in a 37-hectare hyperscale data centre complex in the commune of Hollands Kroon. Forty kilometres north of Amsterdam, this rural sector that has some of the largest greenhouses in Europe was ripe for a development of this kind given the available space and energy, but also because its connectivity from the Agriport site had five optical fibre lines connected to the AMS IX. Today, the first two sectors of the Microsoft campus are completed and the two others are under

[10] The real estate development company has the same name as the sector itself.

[11] The development agency for the region north of Amsterdam (NHN) tries to attract companies by preselecting sites, organizing visits and providing information on tax incentives.

Figure 7.6: Microsoft substation in Agriport

Source: © Fanny Lopez, 2020

construction. In 2018, Google acquired 71 hectares to expand its first data centre that had already been built in Agriport A7, at the same time as the enlargement of another one located in Eemshaven (where a 25-hectare site is already operational).

According to an article in the press in early 2019, CyrusOne was also interested in a 30-hectare site in this same zone.[12] In 2011, a private micro-electric grid was installed, one of the rare ones in the country.[13] Thirty-five gas cogeneration plants are currently connected to the grid with a production capacity of 135 MW, linking greenhouses, logistics spaces, offices and data centres. A heat recovery project is being studied. It is the ECW company that owns and operates a private micro-grid and the infrastructures of the park, notably a substation connected to the TenneT transmission grid with three 80 MW transformers with a total capacity of 240 MW. In October 2010, ECW installed an intelligent grid tool (EWEB). Thanks to this online

[12] 'De Volkskrant Vruchtbare grond voor datacenters: familiebedrijf ziet techgiganten naar de polder komen', 18 January 2019, Available from: https://www.volkskrant.nl/economie/vruchtbare-grond-voor-datacenters-familiebedrijf-ziet-techgiganten-naar-de-polder-komen~b9cd27e1/ [Accessed 5 February 2019].

[13] Digital Gateway to Europe (Stijn Grove, Judith de Lange), *Rapport North Amsterdam Data Center Campus: Economic Impact Study*, June 2018, Available from: http://www.digitalgateway.eu/uploads/1/6/1/8/16187644/northamsterdamdccampus_eng.pdf [Accessed 5 February 2019].

Figure 7.7: Map showing Agriport, with its zones dedicated to data centres in red and greenhouses in yellow

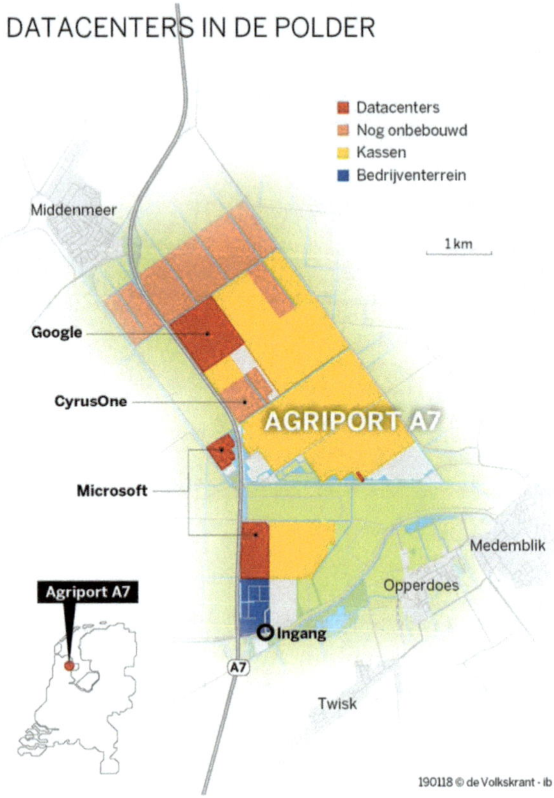

Source: De Volkskrant, January 2019

tool, the different customers can exchange electricity and optimize demand. This has resulted in a significant reduction in transmission costs, which are based on peak demand.[14] In 2017, TenneT started building a second 480 MW substation for later developments.

All of the Netherlands is impacted by hyperscale data centres, unlike Dublin, which kept them in proximity to its metropolitan region. In a highly controlled regional development, these 'rural' territories bring together energy production, agriculture, logistics and the digital in a very

[14] Interview with Jack Kranenburg, Commercial Director of Agriport A7, Agriport, March 2020. See also the report *North Amsterdam Data Center Campus*.

Figure 7.8: The Google data centre in Agriport A7, 2019

Source: © Fanny Lopez, 2020

large-scale industrial planning scheme in which the lines of greenhouses are only equaled by the façades of data centres.

Like Ireland and the Netherlands, Sweden is one of the European meccas of the digital industry market: the contracted has attracted the investments of colocation data centre operators and hyperscale companies. According to Cloudscene, Sweden's total GDP is over 496 billion dollars, with 6.5 per cent of that figure resulting from the digital economy. According to Invest Stockholm and Business Sweden, the Swedish capital concentrates 60 per cent of the Nordic colocation market, thus dominating 'the capacity, networks and Internet traffic with 125 to 150 network service providers active on the market'. It should moreover be noted that 55 per cent of Nordic Internet traffic passes through Stockholm. According to Cisco's 2021 projections, the IP traffic and the Internet traffic in Sweden doubled between 2016 and 2021.[15] A study commissioned by the Nordic Council of Ministers projects strong growth in the data centre market until 2025, with annual construction investments on the order

[15] Sweden – 2021 forecast highlights, CISCO, Available from: https://www.cisco.com/c/dam/m/en_us/solutions/service-provider/vni-forecast-highlights/pdf/Sweden_2021_Forecast_Highlights.pdf [Accessed 5 February 2020].

of €2–4.3 billion.[16] The attraction of the digital sectors can be explained by four reasons: a cold climate that means less electricity consumption for cooling, land availability, the attractive price of electricity combined with low taxes[17] and a large quantity of electricity from renewable and decarbonized energy sources.[18]

Gävle and Sandviken: Swedish hub of GAFAM

The large data centres are primarily installed in the Stockholm region, but also outside the capital, especially in the northern part of the country. The establishment of the hyperscale started in 2011 when Facebook decided to build its first European data centre in Luleå near the Arctic Circle. It is an enormous building 320 metres long, 100 metres wide and 30 metres high.[19] The press largely covered the event, vaunting in particular a 100 per cent renewable electricity consumption thanks to hydroelectricity. Today, Facebook has two large data centres in the region and the construction of a third is under discussion. Other rural regions have seen gotten involved in similar installation projects. According to the economic agency Invest Stockholm, six municipalities in the Stockholm region now have buildings that are partially operational or under construction: Horndal in the Avesta municipality where Google acquired land; Katrineholm, Västerås and Eskilstuna, where Amazon Web Services has set up shop, as well as Gävle and Sandviken where Microsoft installed a site. Another Microsoft site is in the southernmost part of the country, in Staffanstrop, near the Danish border. Some operators developed several sites in the same municipality, which has brought the total number of hyperscale sites in Sweden to nine. Five others are located in the neighboring countries of Denmark and Finland.

[16] Jakob Dybdal Christensen, Jens Therkelsen, Ivo Georgiev and Henrik Sand, *Data Center Opportunities in the Nordics: An Analysis of the Competitive Advantages*, report for the Nordic Council of Ministers, 2018.

[17] In 2017, the Swedish tax agency permitted the data centres to benefit from the same tax reduction as other industries with a high energy consumption, like electricity production plants. The customers consuming over 100 kW of energy through the digital infrastructure can ask for a reduction on the tax on electricity. Between 2017 and 2019, the colocation data centres were also concerned by this advantage before it was reserved for the largest consumers. This intensified investments of large-scale operators such as Google, Amazon, Microsoft and Facebook.

[18] Danny Bradbury, 'Super cool: Arctic data centers aren't just for Facebook', *The Register*, 2016, Available from: https://www.theregister.co.uk/2016/05/12/power_in_a_cold_climate/ [Accessed 1 December 2020].

[19] Asta Vonderau, 'Scaling the Cloud: making state and infrastructure in Sweden', Ethnos, 84(4), 2018, pp 698–718.

At the end of 2017, Microsoft confirmed the installation of three enormous sites in the rural municipalities of Gävle and Sandviken[20] located in the county of Gävleborg, some 170–90 km north of Stockholm (in the centre of the country).[21] According to Klara Wirdby, commercial development director for the municipality of Gävle, these projects were strongly supported by the agency Business Sweden, the regional marketing agency, the Stockholm Business Alliance, as well as Node Pole, which has an online platform bringing together sites for industrial use with high energy consumption.[22] These agencies accompanied in a sort of industrial lobby the data centre operators in choosing the ideal place. After having homed in on the county of Gävleborg on a map, the agencies began to accompany the municipalities (which had no experience in the sector), support that would last for 8 to 10 years. Evaluations of environmental impacts as well as a detailed plan of each site were put in place. The municipalities committed to prioritizing the planning process in record time, 10 months for Gävle and 8 for Sandviken, instead of the usual one to two years. The municipalities generally do not issue any specific recommendations on the architecture or landscaping in zones of an industrial nature. According to the municipality, Microsoft had ambitions that exceeded the community's expertise for this type of site. This example made it possible to measure, a little like in Prineville, the power relationship between these big players and the small municipalities caught up in an industrial project that goes beyond them, but from which they hope to get, in exchange for making communal land available, a few small economic benefits from them. In terms of jobs and vocational training, Microsoft took charge of one part of the training of computer technology technicians on the secondary school level. The company also donated equipment and computers, contributed to different NGOs (sports, local information technology events and others) in the two municipalities and financed digital training in the new companies.

[20] At the moment went Microsoft started looking for land, Facebook was also searching for a site in Gävle and the negotiations went rather far. There wasn't enough available capacity for Microsoft and Facebook's development, both of which were asking for a large amount of power. Working with GRD Vattenfall, plans were drawn up so that the two developments could each obtain half of the capacity already in place, to then simultaneously increase power once the investments in infrastructure were made. In the meantime, Facebook found another site, and after its withdrawal from the project, Microsoft was able to benefit from the energy requested.

[21] The distance between the city centres of the two municipalities (Gävle 103,000 inhabitants, Sandviken 39,000 inhabitants) is about 20 km.

[22] Interview with Klara Wirdby, Commercial Development Director for the municipality of Gävle, September 2020.

Figure 7.9: First building of the Ersbo site, May 2020

Source: © Municipality of Gävle

The first of Microsoft's three sites, Ersbo (Figure 7.9), covers an area of 66 hectares of municipal land. The second, Stackbo (Figure 7.10) (next to the substation of the same name), brings together 150 hectares of a private forest company (it is the largest of the three sites, 10 buildings are planned[23]). The third site (Figure 7.11), Tuna, extends over an area of 64 hectares of municipal land about 1 km from the houses of Sandviken.

Already declared industrial zones in the global plan, Ersbo and Tuna are two rather isolated sites, and their installation did not cause any controversy. It was a little different for the Stackbo site, a forest near a small village. These landscape mutations prompted debates and some opposition from the inhabitants. It was notably the construction of a new substation to complement the existing one that mobilized the population. The Internet giant made construction of this substation a condition of its installation. It is a backup connection because the data centre is not equipped with a diesel generator (in order not to mobilize fossil fuel), which is a world first. The data centre will therefore have two separate connections on the national grid. The two substations will be very close together, less than 1 km apart. According to the regulation of the electricity market, concessions for new power lines and the substation are necessary, which may take several years. Vattenfall, the Swedish public electricity production and distribution company, will try however to speed up the construction process.

23 Interview with Klara Wirdby, September 2020.

Figure 7.10: Stackbo, Gävle site under construction, May 2020

Source: © Municipality of Gävle

Figure 7.11: The first building of the Tuna site in Sandviken, 2020

Source: © Municipality of Sandviken

Each site can be divided into five halls or large rooms, which will bring the total number of the buildings to 15 for the two municipalities. The building delivered in Ersbo is 400 metres long and 65 metres wide (26,000 m^2); the construction of the second building started in the fall of 2020. The Stackbo and Tuna buildings were delivered in the spring of 2021. In Stackbo, another substation project is underway. Five substations will be built for the totality of the sites.

With this type of installation, the municipalities, with the regions, have to be able to accompany and facilitate the operators' installation, because each project is very specific. In Gävle and Sandviken, the municipalities as well as the county's administrative council made it possible for an 'accelerated detailed planning' and 'urbanism procedures to be well-aligned on Microsoft's requirements'.[24] Here, as elsewhere, the public authorities served private companies in a worldwide race to monopolize land and electricity because this is what it is a question of.

[24] And not only detailed planning but the process as a whole with the possibility of giving Microsoft direct answers on what was possible and what wasn't. A large part of this process came from the close collaboration between the Swedish actors (BS, SBA/IS, CAB and municipalities).

8

Growth, Acceleration and Digital Productivism

Steered by the University of California, Berkeley, the study *Datagravity*[1] provides information on how Forbes' 2,000 largest worldwide companies exchange data. One of the results is acceleration and densification in a 'snowball effect' around 50 or so cities in the world. The data goes where the data is.[2] There is a major data concentration and centripetal aggregation dynamic in proportions that create perplexity. This phenomenon does not mean that there are no data centres in the region; there are, more and more, and they are getting larger. However, these are not the same data; it is mostly about distribution and collection.

The continuation of the digital growth dynamic is explained by the evolutions in the technologies deployed (optronics, wavelength, etc), but also by the creation in 2022 of new undersea optical fibre cables: the Peace, 2Africa, Ellalink, Maréa, Dunan and Amitié cables. The Chinese Peace cable[3] linking Pakistan to the Prado beaches in Marseille in the south of France represents, with these 320 terabytes[4] per second, itself, the double of the 15 undersea cables connected in the large Marseille data centres. The 2Africa cable rolled out by Facebook will also revolutionize African connectivity and the interconnection with the Middle East and Europe. Over 45,000 km long, it has become the world's longest, linking three continents, almost the equivalent of the Earth's circumference. This 160-terabyte 2Africa cable (upgradable to 320 terabytes) will open the 4G and 5G channels. It literally goes completely around the African continent

1 Digital Realty, *Data Gravity Index DGx, Report Measuring the Intensity of Data Gravity and Its Effect on the Global 2000 Enterprises*, 2020.

2 Fanny Lopez, *A bout de Flux*, Paris, Éditions Divergences, 2022.

3 Peace for Pakistan and East Africa Connecting Europe.

4 A terabyte is a measurement or storage unit in computer language. It is equivalent to a thousand gigabytes, or a million megabytes, or 1 trillion bytes.

and all the neighbouring countries as far as Saudi Arabia, coming back to interconnect in Marseille, which confirms its place as a worldwide gateway. Among the other announcements that have excited the digital world is the Ellalink cable. It is the first undersea cable in history to interconnect South America and Europe without passing through the United States. It starts in Brazil, in São Paolo, passes through Rio, Fortaleza and Lisbon, and will interconnect Madrid, then Marseille.[5] Connectivity is strengthened even further with the installation of three other major cables in 2022: the Maréa cable (Facebook[6]), Dunan (Google)[7] and Amitié (Facebook).[8] Each 160-terabyte cable brings the American transatlantic flow directly to the western side of Europe, on the Bilbao side and the Gulf of Gascogne in France, and no longer via the north, which has traditionally been the case since the invention of the telegraph.

In 20 years, the cost of the cable has fallen to a tenth of what it was, and its capacity has increased by 50-fold. The decrease in the cost of a telecom link is on the order of a factor of 500. And it is not only the cables that are multiplying with faster optical fibre, but the wavelength at the end of the cables will also go ever further. The switch from an underside optical fibre to a terrestrial one opens a range of telecom links called wavelengths, which is what makes it possible to access the community of customers (operator, CDN,[9] Cloud actor, etc). Until today, there had been a physical constraint of 50 km between the landing point and the installation point of the equipment that creates the wavelengths for 1, 10 or 400-gigabyte links. It is this equipment, also called Submarine Line Terminal Equipment (SLTE), made by Nokia, Huawei, Alcatel and others, that creates the actual length of the cable. The cable, however, is only a conduit; it is the wavelength at the end of the cable that has value – it is here that the flow space resides. Between the landing point and the location of the SLTE, the necessity to route the cable around buildings

[5] This is also without taking into account the American flows that traditionally went through Nova Scotia, Ireland and England, to then go south to Europe.

[6] The Maréa cable, built by Facebook and Microsoft, leaves Virginia and arrives in Bilbao, then continues underground to Madrid and Marseille.

[7] The Dunan cable, which is Google's 14th undersea cable, links Virginia Beach, in the United States, to Saint-Hilaire-de-Riez, in Vendée, in France. This telecom line creation equipment is not in Vendée, but in the Interxion data centres in Paris. The geography of data shortens distances: Paris is in Vendée or Vendée is in Aubervilliers.

[8] The Amitié cable initiated by Facebook, which leaves the state of Massachusetts (in the United States), arrives off the shore of Le Porge, on the Aquitaine coast in France, and continues to Bude (in England). In France, this cable is connected to the data centres of Equinix's transmission facilities in Bordeaux, before redirecting the flow in a Y: to Paris on one side and Marseille on the other.

[9] Content Delivery Network.

reduced the constraint from 50 km to 25 km in urban zones like Paris and Marseille. These last few years, a major leap has been made by the optical fibre designers, increasing the possible distance between the landing point and the SLTE equipment from 50 km to 500–600 km. Whereas the challenge for a flow operator is to make the cable arrive where the telecom flow is consumed – that is, in hubs – the increase in this distance enables the cables to go directly to the data centres to find their market through clientele concentration and vice versa. This technological progress has favoured the over-concentration of flows phenomenon in the worldwide hubs. Unlike other companies, the data centre cannot relocate its raw material – the data flow, whose mobilization and interconnection speed comprise its value.

One of the spatial entities that shows this accumulation is Ashburn, Virginia, 48 km northeast of the US capital, Washington, DC. It is the largest data centre concentration worldwide: a sort of large suburb where around 300 data centres consume 3 GW, or the equivalent of two nuclear power plants. The GAFAM companies (Google, Amazon, Facebook, Apple and Microsoft) are present there, with their own data centres, but there are also the large colocation operators[10] like Equinix, Interxion, Digital Realty and many others. For digital industrialists, Ashburn foreshadows an international trend: hyper-concentration.[11]

This has a rebound effect, because the growth and acceleration of flows and interconnection in urban hubs favour at the same time the growth of storage and redundancy data centres in the middle of the countryside. The business world gravitates towards the hubs of certain cities and certain colocation data centres that dominate the worldwide market like Interxion, Digital Realty and Equinix. The GAFAM companies need these service providers that accompany them from Tokyo to Los Angeles by way of Paris and Frankfurt. In a single order, one of the Big Tech companies reserved 10,000 m^2 in the new Interxion data centre in La Courneuve. Today, the Plaine Commune digital hub is at a saturation

[10] Data centres can be divided into two large types: enterprise data centres and colocation data centres. Enterprise or administration data centres host and manage their own data on their servers in a building that belongs to them. Colocation data centres have more varied uses: hosting client companies' digital equipment (the operator provides the space and electricity); the availability of the host's computer servers and equipment for its customers (the customers can make temporary physical reservations of the server, disk racks and network equipment in order to benefit from a guaranteed and unshared use of the infrastructures); reservations in Cloud mode (the customers can reserve virtual machines on the data centres' servers).

[11] Interview with Fabrice Cocquio, President of Interxion France, September 2021; interview with Régis Castagné, President of Equinix France, October 2020.

point and the creation of a new 1 GW site is under discussion on the scale of the metropolis. This represents half the size of Ashburn data centre concentration. In order to reduce conflicts of use, all the large digital metropolises are reflecting on digital zones for an exclusive use of the data storage function.

PART III

Energy Disturbances and Conflicts of Use

A new spatial entity, data centres have consolidated the territories on a continuous digital landscape. Whereas the habitability of dense urban zones depends increasingly on their digital networks, data centres appear as a new piece of the 'energy landscape' linking the national or even transnational scale to the local scale. The recent debates and publications on the energy transition and the adaptation of cities to climate change have led to rethinking the role of energy production systems, notably in environments with a high population density.[1] The relationship of dependence of urban centres vis-à-vis the worldwide 'energy hinterland'[2] has undergone modifications over the last 15 years. The search for and integration of more decentralized energy production systems have hybridized the traditional functioning of the electricity grids. Densely populated zones are becoming increasingly involved in producing renewable energy sources, through integrated micro-electricity production plants, but also micro-grids, which is notably the case in New York and London.[3] The relationship of interdependence between architecture, infrastructures and the territory in a world with limited resources has forced reconsiderations and redirections that cannot be limited to questioning how digital infrastructures can be better integrated and pooled. The debate on the ideal digital society is not the object of this

1 See notably the colloquium: Istituto di studi urbani e del paesaggio (ISUP), *More or Less Density*, organized by Jonathan Sergison, Frédéric Bonnet, João Nunes, Sascha Roesler and Enrico Sassi, Università della Svizzera Italiana, Mendrisio, 29–30 October 2020.

2 Dean Hawkes, *The Environmental Tradition: Studies in the Architecture of Environment*, Oxford, Taylor & Francis, 1996.

3 GLA, *Decentralised Energy Master Planning: A Manual for Local Authorities*, 2011. See on this subject Fanny Lopez, 'Keep the lights on! La décentralisation énergétique à Londres', in Gilles Lepesant (ed), *Énergies nouvelles, terriroires autonomes?*, Paris, Presses de l'Inalco, 2018.

work, which has limited itself to giving an account of data centres' spatial and energy impact as revelatory of an untenable situation over the long term.

The energy density of data centres has no equivalent in the history of urbanism. These centres disturb local energy systems and their accumulation in urban zones in the same way as their extension into rural territories is a great concern for planning on every scale. The destabilization they cause in the local electricity systems has given rise to very different reactions. The responses of the public authorities of certain European cities like Amsterdam and Stockholm have received a great deal of attention for the associations between urban and electricity planning actors, but also for new energy pooling based on heat recovery like the Data Parks program of Stockholm. In this part, we will address the prospects of ecosystemic exchanges. If data centre infrastructural reserves are very substantial, successful pooling is rarer because it requires, very much upstream, an association of the chain of energy and urban actors, and the appropriate economic and legal model.

9

Rising Electricity Consumption and Territorial Destabilization

Despite the improvement in the energy efficiency of the components, the global electricity consumption of data centres has substantially increased. In academic and industrial research, the energy efficiency of the digital in its use phase is a fully-fledged research topic. In 1965, the cofounder of Intel, Gordon Moore, established in an article that the power of information systems doubles every 18 months. More recently, at Stanford University, Jonathan Koomey and his colleagues demonstrated that the quantity of energy that a machine needs to carry out a given number of calculations decreases by a factor of two every year and a half.[1] However, energy dissipation (notably thermal) and also the multiplication of machines keep growing, as anticipated by the physicist Rolf Landauer[2] in the 1970s. Research on the energy impact linked to the computing of frugal data centres started in the early 2000s with the work of Wu Feng and his team at Virginia Tech. In the framework of the Super Computing in Small Spaces project, the researchers developed Green Destiny: Low-Power Supercomputer (2001), which succeeded in putting supercomputers into production in reducing costs, the heat emitted, the energy consumed and the space required. In France, at the INRIA, the EcoInfo group studied the impacts and rebound effects of these different phases.[3] As for the buildings themselves, new constructions are generally very energy-efficient as the data centre operators have the financial resources to

1 The limits seemed to have been reached today; the theoreticians of quantum mechanics have warned that we won't be able to store more than 1 byte of information per atom.

2 Rolf Landauer, 'Irreversibility and heat generation in the computing process', *IBM Journal of Research and Development*, 5(3), 1961, pp 183–91: 'No matter what irreversible logical manipulation of information, such as the elimination of a byte or the merger of two calculation paths, is accompanied by an increase in entropy in degrees of the non-informational freedom of the information treatment device or its environment.'

3 Groupe EcoInfo, *Impacts écologiques des Technologies de l'Information et de la Communication – Les faces cachées de l'immatérialité*, Les Ulis, EDP Sciences, 2012.

Figure 9.1: Share of communication technology of global electricity usage

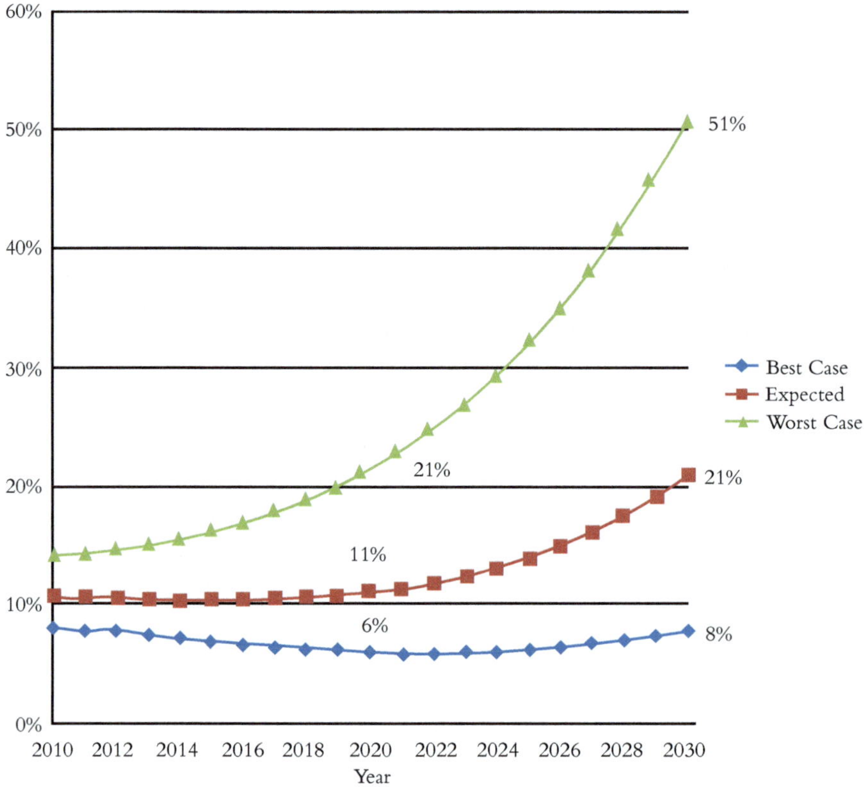

Source: Anders S. G. Andrae and Tomas Edler, 'On global electricity usage of communication technology: trends to 2030', *Challenges*, 6, 2015, pp 117–57

invest in very upmarket equipment (undulators, batteries, air-conditioners, controllers, etc). Despite these efforts on the buildings' energy efficiency as well as the innovation and acceleration prospects (computing power, transmission speed) of software and hardware, the global energy consumption of infrastructures dedicated to data processing is rising at a dizzying rate. According to a reference study undertaken in 2015 by Anders Andrae and Tomas Edler of the Huawei R&D centre in Stockholm, the digital sector consumed 7 per cent of worldwide electricity in 2013.

Data centres themselves represented 2 per cent of the worldwide total.[4] Their projections reach a maximum of 13 per cent of worldwide electricity

[4] Andrae Anders S. G. and Edler Tomas, 'On global electricity usage of communication technology: trends to 2030', *Challenges*, 6, 2015, pp 117–57. In this study, electricity consumption in 2013 was estimated at 21,000 TWh (7 per cent represents 1,470 TWh).

consumed by data centres in 2030, and 51 per cent for the information sector in its totality. In 2017, the négaWatt association estimated consumption in France at 9 per cent.[5] The think-tank The Shift Project recently revised this disaster scenario downwards, but nevertheless estimates that the digital sector could represent 25 per cent of worldwide electricity in 2025.[6] For 2030, the large metropolises anticipate an increase in electrical power of at least 1 or 2 GW on average, but this could be greater.[7] Let us recall that 1 GW is the average electrical power of a reactor in a modern nuclear power plant. The digital technical system follows an exponential curve in terabytes and megawatts consumed. Twenty years ago, the first data centres had on average 1,000 m^2 of computer rooms and functioned with 1 or 2 MW. Today, certain large sites, like the new Interxion campus in La Courneuve, offer some 40,000 m^2 of space and show a consumption of 130 MW. The major CEOs of the sector, in what can resemble a self-fulfilling prophecy, envisage that within five years a large data centre will have no less than 60,000 or 100,000 m^2 in rooms and some 300 MW in electrical power.

The entire electricity sector is disturbed by the digital. Apart from transmission and distribution, it is also the production sector and this holds true throughout the world. The GAFAM companies (Google, Amazon, Facebook, Apple and Microsoft) exert pressure on the energy production of the historic electricity companies, getting involved occasionally in the massive co-financing of renewable energy production infrastructures. For Gary Cook of Greenpeace, author of the Click Clean report,[8] if the GAFAM companies make an effort, notably in showing commitments to being supplied by 100 per cent renewable energies, there is a long road ahead because this energy is neither local, nor additional, nor really used by data centres: it is partially purchased by electricity companies by means of certificates (RECs in the United States, GO in Europe). The GAFAM companies are increasingly vigilant and demanding and go as far as forcing the energy transition of certain old utilities to use a very carbonized mix … This is notably the

---

and the projections for 2030 reached 61,000 TWh. A nuclear reactor produces 7 TWh annually.

5 Association négaWatt, *Scénario négaWatt 2017–2050*, 2017, p 8, Available from: https://www.negawatt.org/IMG/pdf/synthese_scenario-negawatt_2017-2050.pdf [Accessed 2 March 2018].

6 Hugues Ferreboeuf (ed), *Lean ICT, Pour une sobriété numérique*, The Shift Project report, October 2018.

7 Interview with Enedis and RTE, February 2020. See also The Data Centre Reports, DCByte, Available from: https://www.dcbyte.com/knight-frank-data-centres-report/Q2-2020 [Accessed 2 February 2020].

8 Interview with Gary Cook of Greenpeace, San Francisco, September 2017. See also the report, *Click Clean Virginia*, Greenpeace, Available from: http://www.greenpeace.org/usa/global-warming/click-clean/ [Accessed 2 September 2017].

case in certain US states where land is inexpensive but the energy mix has a large amount of fossil fuel in it (with coal plants for which the industrial lobby remains very powerful). As highlighted in the 2019 ADEME report, certain GAFAM companies develop their own production infrastructures on site or nearby. In Oregon, Apple bought the 45-mile hydroelectric dam to supply its Prineville data centre (Figure 9.2). In Wyoming, Microsoft purchased over 230 MW of wind-powered electricity from the local historic electricity company. In Virginia, Microsoft had negotiated an agreement with Dominion Virginia Power and the state government to take over the production of a 20 MW solar energy plant. The company has done the same thing in Europe, where it signed a contract to reserve the entire electricity production of a General Electric wind farm, which had just gone online in Kerry County in southwestern Ireland. This purchase aims to 'green' the energy mix of its large data centre west of Dublin dedicated to Cloud services for its European corporate clientele.

These examples are in no way anecdotal. Not only does digital infrastructure influence electricity production, but it also equally disturbs transmission and distribution because it is supplied by connecting to the historic electricity grids. A data centre generally benefits from several connections to these grids, two on average, but also from an onsite electricity autonomy through a micro-grid that interconnects backup electricity production (from a fuel or gas combustion motor) and a battery storage system. Ensuring electricity

Figure 9.2: Power lines in Prineville, 2017

Source: © Fanny Lopez, 2017

continuity is the first obligation of a data centre and this is where its value lies. This equipment[9] operates on average three hours a year, half for tests. This backup is not intended for internal breakdowns as, this system's security is guaranteed. Its purpose is to offset any possible failures on the public distribution grid, which is considered fragile due to a lack of investment, especially in the United States.

The operators of the electricity distribution and transmission grids are in constant contact with this industrial clientele whose requests and requirements continuously increase. Over the last ten years, digital installations have been mostly created on a case-by-case basis, without any global and strategic vision. But today, heavy pressure and destabilizations that make data centres weigh on electricity systems have forced these systems to move a subject that has remained relatively opaque at the top of their policy agenda. In the United States, as in Europe, electricity operators are rather discreet, given the fact that keeping the companies' confidentiality is mandatory. But exchanges with the sector's principal actors show difficulties that are increasingly less manageable internally. It is the increase in data centres' electricity consumption , first and foremost, that is threatening to create, in places, capacity shortages on the electricity grid. These are often caused by a difference between the data centres' planning horizon and that of the transmission and distribution operators. Data centres usually 'over-reserve', that is, they contract for much more electrical power than they generally need, notably to curb the competition. Their demand is often based on a maximum consumption hypothesis (when the data centre will be full – which sometimes takes several years). This gap between the reserved and the consumed power blocks other customers on the territory. Let us recall that the principle of an electrical connection is the same in most European countries: it is the 'first arrived/first served' principle. Whether it is a data centre, a hospital or a housing complex, the customer asks the public service distribution grid for a certain number of MW, and it is obliged to supply the electricity requested. In dense locations, this over-reservation creates conflicts of use for other urban developments. In Île-de-France, Enedis has had difficulties in supplying certain territories like Plaine Commune with electricity, which creates tension in the communities. Another example is Marseille, in the south of France: a data centre reserved a large quantity of electricity (90 MW), leaving nothing for the city's electric buses. The mayor of Marseille had to negotiate to recuperate the missing 7 MW for the city's fleet.[10] The increase in data centres' electricity consumption has also

[9] The Interxion data centre on the rue Rateau has 14 4 MW motors for a total of 46 MW and three 80,000-litre tanks for fuel oil.

[10] Interview with Fabrice Cocquio, President of Interxion France, Paris, March 2018.

caused an unbalanced division of infrastructure costs, whereas the urgency of the energy transition has simultaneously led to large costs to strengthen the electricity grid. Imbalances have appeared between the development of this private industry and the preservation of community interests, of which the public operators are the guarantor. Faced with the growth in calls for electricity consumption, the grid's managers are looking for solutions to offset the risks of overinvestments linked to the power reserved by the data centres. In every city in which the data centre industry is deployed, the electricity operators have experienced tension in terms of their grid and are trying to work more closely with urbanism departments to set up a spatially and energy-controlled siting strategy in the medium and long term.

10

Île-de-France in the Quest for a Strategy

The geography of data centres in Île-de-France was first concentrated on the Plaine Commune territory before becoming more complex over the last few years (Figure 10.1). Emerging clusters have thus been confirmed. The Velizy-Villacoublay activity zone, south of the Meudon forest and alongside the A86 highway, consequently now has an increasing number of infrastructures, while a growing dense string of installations can be found along the A4 highway, running east from Paris to Strasbourg (enterprise data centres). Each time, the presence of strong connectivity along highway infrastructures and land availability play a major role. In the south of the Paris region, the Saclay/Orsay hub is also incredibly attractive due to the availability of electricity and land. The Data 4 campus has consequently increasingly densified, like the Courtabœuf activity zone in Orsay. In the western part of the region, the enlarged La Défense sector is also being maintained, due to the presence of the head offices of many large international companies. The regional digital geography is therefore being transformed, vis-à-vis the complexification of the data centre system (installations according to uses, various building sizes, strong growth and extension needs).

In the Île-de-France region, between 2019 and 2024, Enedis and RTE[1] observed a growth in requests for connection studies from data centre

1 Formerly brought together in the EDF group, the liberalization of the energy market marked the separation of the distribution (Enedis), transmission (RTE) and production (EDF) branches. Enedis was tasked with the management and development of 95 per cent of the electricity distribution grid in France. Over 80 per cent of EDF was held by the state. RTE by EDF: 50.1 per cent; Caisse des dépôts et consignations: 29.9 per cent; CNP Assurances: 20 per cent; Enedis: 100 per cent subsidiary of EDF. Despite some divergences and upcoming liberalization prospects, the interdependence of these three branches remains very strong and transmission and distribution work hand in hand to optimize the major networking on the national scale and these European interconnections.

Figure 10.1: Map of data centres in the Paris region, 2021

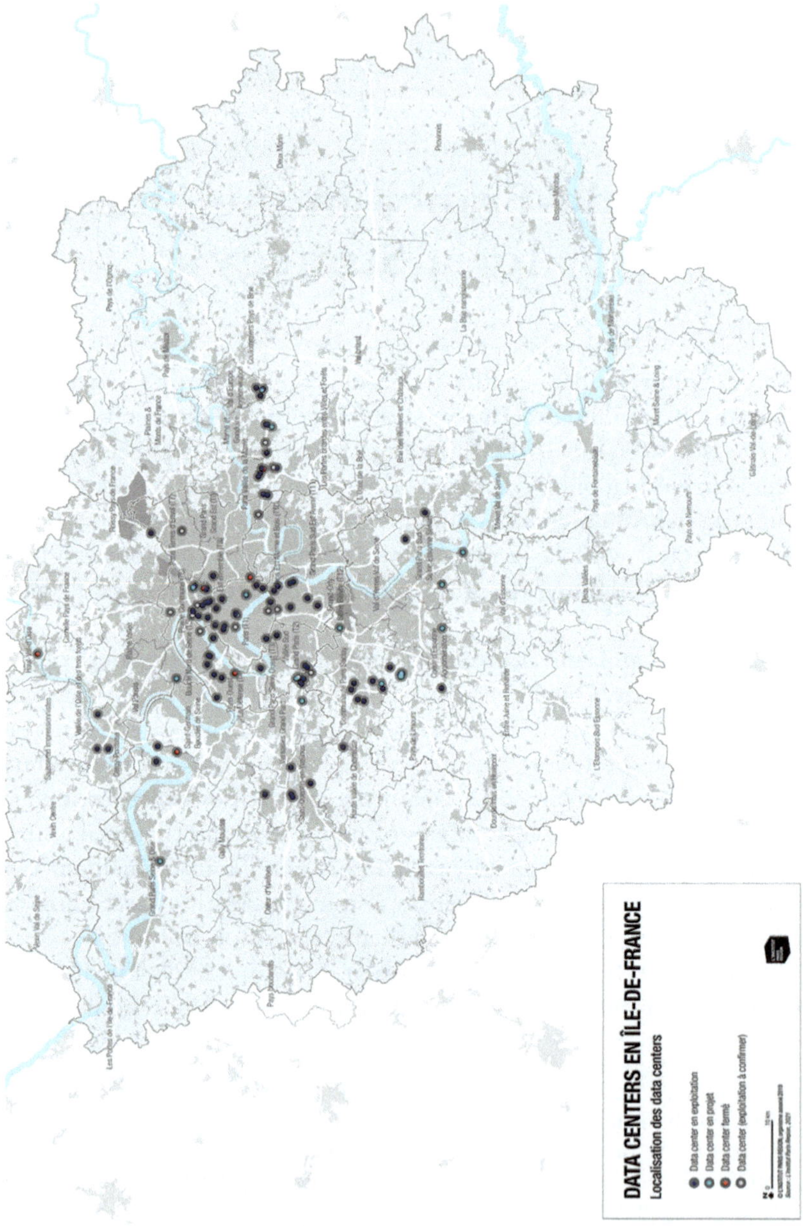

Source: © Institut Paris Region

operators. Even back in 2012 – when the subject remained marginal in the field of public policies – the regional climate, air and energy plan (SRCAE), adopted in December of that year, showed data centres as a 'new potential', a 'possible source' in the framework of developing the heat recovery energy sector. As for energy consumption, the text pointed out that a 10,000 m^2 data centre needs a 20 MW connection to the electricity grid and consumes as much electricity as the average-sized city of 50,000 habitants.[2] The recovery of heat emitted by the data centres was perceived as a new challenge specific to the Paris region, but would warrant 'being better evaluated in terms of potential'. The installation of heat recovery at Val d'Europe in Seine-et-Marne (under construction at the time) was cited as an example. The territorial administrations were recommended to 'contact the existing data centre operators to study the possibilities of recovering and using waste heat (connection to a heating network, heating for buildings, swimming pools, hospitals)'. After the SRCAE, in 2013, the reports of the Regional and Interdepartmental Division of the Environment and Energy (DRIEE) and the Local Energy and Climate Agency of Plaine Commune drew attention to the impact of the digital. In its study on the new consumptions of Grand Paris, the DRIEE announced, for electricity, a call for additional power of 4,000 MW divided as follows: 400 MW for the 72 metro stations of the Grand Paris Express, 800 MW for the 800,000 additional housing units, 1,300 MW for 1 million jobs in the service and industrial sectors, 500 MW for 1 million electric vehicles and 1,000 MW for data centres. In this study, data centres represented a quarter of additional needs in electricity by 2030. These were largely cited in the press because they showed an increase vis-à-vis the SRCAE trend of over 2,200 MW, whereas controlling electricity consumption was recommended. In June 2024, on the basis of currently approved planning authorizations, the connection power for RTE was around 3.5 GW. This figure does not include the pre-study requests or the requests for connection on which no further action is being taken for the moment.[3] In France, the Enedis average-voltage distribution grid can supply up to 40 MW of electrical power. Either the data centre connects to an existing substation, or Enedis builds one to make the electricity requested available. If a data centre asks for a connection above 40 MW, it is directly connected to the RTE high-voltage transmission grid. For colocation data centres, the requests are generally on average between 5 and 20 MW. At

---

[2] Région Île-de-France, Schéma Régional du Climat, de l'Air et de l'Energie de l'Île-de-France, 2018, Available from: https://www.drieat.ile-de-france.developpement-durable.gouv.fr/IMG/pdf/14-SRCAE-IDF_Synthese_actions_recommandees_collectivites_cle4114e6_cle18d814.pdf [Accessed 14 March 2025].

[3] Aggregated figures provided by RTE and Enedis during a work session in 2020, non-public data.

40 MW, the customer can choose between Enedis and RTE. An Enedis connection is often faster and easier because it does not require a particular authorization; the work takes less time, between two and three, sometimes four years. At RTE, the time frame can be as long as five to seven years. For a customer's first connection, the discount rate applied by Enedis is 40 per cent, while at RTE, it is only 30 per cent. This therefore leads to a tendency to facilitate connections to distribution grids. Historically, and it is still the case today, Enedis received more connections than RTE. Over the long run, the cost of access to the network is lower at RTE than at Enedis. The cost of energy over the long term is more advantageous on the high-voltage grid because it is 'big', but the data centre operator must acquire expertise to be able to operate a substation and it must also partially finance it. Over the last few years, the major tendency to have a connection at Enedis has been modified. Large-scale data centres are multiplying and an ever-higher number of requests for very high power (sometimes above 100 MW) is now made directly on the RTE transmission grid. The technical specificities of the transmission grid and that of distribution are not the same. RTE's transmission grid capacities make it possible to have a greater margin and there is generally more available capacity. This is much more difficult for Enedis, which has more complex grids with a tighter grid interconnection. However, even on a very well-linked transmission grid like the one in Île-de-France, the multiplication of these high capacities creates problems because adjustments and reinforcements must be made.

In this context, electricity distribution operators are debating the questioning of the historic founding principles and the related regulations. Whereas the growing scarcity of resources makes it necessary to think about treating with caution, it is no longer taboo for distribution operators to bring up regulatory evolutions in order to limit over-reservation. A discussion notably took place in February 2022 at the Dutch Parliament and an amendment was tabled so that the connection on the 'first come, first served' and 'without any conditions' principle be reasserted as an unchangeable one. The amendment was not put to the vote. Reflections were underway in France in 2021. Enedis has since set up a flexible demand policy like the one in Ireland. This regulatory measure makes it possible to limit the difference between the connection power requested and the power finally subscribed, in order to avoid over-investments on the grids for power requests that will never materialize. A deadline is therefore set to reach the consumption requested. Today, one of the challenges data centre development is to coordinate spatial and energy planning.

The grid requires strong reinforcements and new electricity sources. RTE confirmed this tendency in the first few minutes of a large press conference in 2021 that announced the results of its prospective exercise on the '2050 Energy Futures': 'We must produce more electricity because we are going to consume more.' RTE projected a threefold increase in data centre consumption by 2050. This increase is part of a phenomenon of global growth of electrification anticipated by the grid operators over the last several years. One of the challenges of energy policies being the electrification of uses, an adaptation of production capacities and a reinforcement of the electricity system must be imposed. This increase concerns many other sectors. The electrification of uses means a transfer of uses that once depended on oil and gas to electricity. It is, for example, the conversion of blast furnaces into electric ones or the shift of internal combustion vehicles to electric vehicles. Electricity will become the main source of energy in France, replacing oil with the development of new-generation nuclear power and/or a strong development of renewable energies (notably large wind and solar farms) – which have become competitive, according to the Executive Director of RTE. In these '2050 Energy Futures', there are no degrowth scenarios, but a downward energy control illusion was maintained throughout the conference, introduced by a paradoxical sentence that acted as a smokescreen: 'Energy consumption will decrease just like CO_2 emissions, but that of electricity will increase.'[4] Yet increasing the electrification of the energy system to lower the carbon footprint in no way guarantees any modification like a radical change in lifestyles, consumption and production towards restraint, the only way to reverse the trend towards a collapse of the Earth system.[5] No structural redirection was announced and, in the end, it is an increase in production that has occurred. It is the same rhetorical motivation that consists in announcing that the consumption of data centres will be offset by the improvement in energy efficiency in other uses, whereas the reference scenario anticipates a threefold increase of this industry by 2050.[6]

RTE and Enedis are both favourable to creating 1 GW[7] clusters in Île-de-France in order to create dedicated zones like Ashburn and to avoid saturations in other territories. The prospective discussions remain very uncertain, just like the measures taken by certain municipalities like Stockholm, with its data parks, and Portland, with its backup generator network. The idea that data centres can contribute to balancing the electricity system, that their

4 Thomas Veyrenc, Executive Director of RTE and Xavier Piechaczyk, Chairman of the Board of RTE, RTE press conference, 'Futurs énergétiques 2050', Paris, 25 October 2021.

5 See IPCC report, Climate change 2021, https://www.ipcc.ch/report/ar6/wg1/downloads/report/IPCC_AR6_WGI_SPM_final.pdf.

6 'Futurs énergétiques 2050: principaux resultats', October 2021, Available from: https://assets.rte-france.com/prod/public/2021-10/BP2050_rapport-complet_chapitre3_consommation_0.pdf [Accessed 25 February 2025].

7 Reflections in the framework of the working group on the regional strategy for installing data centres led by the Institut Paris Region, with Arec, RTE and Enedis.

backup systems (generators and batteries) can be further called on, is being promoted. This could be achieved by: (1) disconnecting the data centre from the grid and using its own backup energy to relieve the grid (provided that the backup systems do not create pollution); and (2) using their batteries to produce and store electricity to complement the grid.

Many pooling prospects are being raised (heat networks, sharing of backup installations, etc), but implementation often remains very complex. Data centres are above all an interconnection and storage industry, and implementation time frames like investments and technical blockages are often obstacles.

11

Infrastructural Redundancies and Difficult Pooling

Data centres are buildings working like machines where everything has a backup system so that they will not be weakened by possible power cuts that could affect the network and impact storage and, in the end, lose market shares. If the risk related to disasters varies according to the regions, all the data centres are extremely concerned about guaranteeing their customers' energy autonomy in the event of problems, which is a central sales argument. Power cuts that last several hours remain an exception, however, such as Hurricane Sandy, in 2012, which plunged one-third of the New York population into the dark and caused the breakdown of 50 per cent of all data centres even if the backup systems enabled, in certain cases, service continuity. Data storage rooms are doubled by infrastructural reserves in adjoining spaces, on the same level, but also sometimes in the basement, on other floors or on the roof. This infrastructural multiplication is a track for pooling that could permit offsetting the impact of their electricity consumption, but exchanges with other actors on the territory and a technical and financial balance remain complicated to find.

Sharing technical equipment?

In New York, data centres present extremely dense technical roofs. They are dizzying outgrowths, including diesel-fired backup generators, water cooling towers for air-conditioning, satellite antennas of every size, water tanks and even cranes to hoist diesel generators up from the street (Figure 11.1). Their basements are filled with cables (Figures 11.2 and 11.3), equipped with fuel oil tanks of several hundred thousand litres to supply the generators in the event of a shutdown of the electricity grid by Con Edison, the main utility serving the city. This is notably the case of 111 8th Avenue, where fuel oil reserves of 600,000 litres can take over energy supply in the event of a power cut. Google and its service manager Taconic carried out a feasibility study

Figure 11.1: Roof of 111 8th Avenue

Source: © Fanny Lopez, 2018

for the construction of a gas cogeneration plant, but have still not decided on this option. They are studying the possibility today of recycling organic waste produced in the building to turn it into energy and use biogas or methane as the primary energy to replace the fuel oil generators.[1] In terms of heat reduction, water cooled at night (when electricity is inexpensive) in tanks is restored during the day for air-conditioning the data centres (Figure 11.4).

The Sabey data centre, previously cited, has similar characteristics. The building has reserves of 1 million litres of water on the ground floor, and fuel oil storage of 700,000 litres (or autonomy at a full load of 72 hours). This infrastructural redundancy raises in the first instance the question of optimization of resources and onsite systems. For example, instead of thousands of litres of fuel oil stored for the backup generators, gas or biogas co- or tri-generation plants appear to be increasingly popular alternative backup systems, and could be a less polluting and less risky option. This is notably the case for *The New York Times* building in New York, whose data centre has a gas cogeneration plant as a backup system. This is also the case for the Citibank data centre in London, and there are other similar examples

[1] Interview with the energy system managers of the Google data centre, New York, January 2018.

Figure 11.2: Access corridor for the data storage room, 111 8th Avenue, New York

Source: © Fanny Lopez, 2018

in Europe. In France, GRDF is diligently trying to develop this market, but the data centre operators have remained hesitant about changing their perfectly mastered system. There is equally a lack of political support for this source, as Guillaume Planchot of the Idex energy company underlines:

> Today, cogeneration in France, with perhaps self-consumption, would be extremely relevant vis-à-vis data centres, even perhaps the

Figure 11.3: Heat conduits, hot and cold water, 111 8th Avenue, New York

Source: © Fanny Lopez, 2018

tri-generation plant. As much as 5,000 MW of cogeneration were installed in France, the hour of glory must have been 2005 to 2007. This equals nonetheless five nuclear reactors. But I believe that over 3,000 were dismounted since there were no longer any purchase rates. They have largely deteriorated, the sector was not supported.[2]

[2] Interview with Guillaume Planchot, Heating and Cooling Network Director at Idex, March 2018.

Figure 11.4: Fuel oil tank and heat storage

Source: © Fanny Lopez, 2018

Fuel cells are also an alternative and certain GAFAM companies are equipped with them.

Following the damage caused by Hurricane Sandy in 2012, micro-smart grids became a central concern for the New York energy actors, financially supported by the New York State Department of Energy. They were perceived as one of the most efficient ways to lighten New York's distribution grids, infamous for their weakness. It was also a way to create resilience in the event of a power cut[3] and create new service loops. The installation of gas cogeneration plants (CHP, combined heat and power) since 2012 in the Framework of the CHP Acceleration Program set up by the state[4] has

---

3 Dan Ton, *Microgrid R&D Program in the U.S. DOE*, November 2018, Available from: https://www.energy.gov/sites/prod/files/2018/12/f58/remote-microgrids-dan-ton.pdf [Accessed 20 March 2021]; Navigant Research, *New York State Microgrid Inventory*, Reference Report No 173511, 13 August 2015 [Consulted online 18 January 2018]; Michele Bebrin, Kenneth Horne and Jonathan Strahl (eds), *Community Microgrid Case Study and Analysis Report*, report for the New York State Smart Grid Consortium, New York, Navigant Consulting, 2015, pp 37, 40.

4 Interview with Ben Falber, energy storage project manager, New York State Energy Research and Development Authority (NYSERDA), New York, January 2018. See also the NYSERDA program.

accelerated: 625 cogeneration plants have been built in the state of New York and 240 in New York City. They are resources that can be mobilized by mini-grids that sometimes include data centres. *The New York Times* Building is a 52-storey skyscraper delivered in 2007 and designed by Renzo Piano.

A nano-grid was installed on the building's scale, partially supplied by a cogeneration system designed by the WSP Flack & Kurtz company. Two natural gas generators supply 1.4 MW of continuous electricity to the building's data centre. The specificity of this installation is that it is continually used as the principal source of energy. Diesel engines and the distribution grid provide backup and redundancy for this system. Water heated by the engine provides heating in winter and also creates an absorption chiller in summer.

There are no lack of pooling examples on urban and regional scales. This is the case of Microsoft and the Black Hills electricity operator, which installed shared gas backup generators in Cheyenne, Wyoming. This installation used as a priority by the Microsoft data centre is also available for the community during peaks on the grid. In 2012, when Microsoft began to negotiate with Black Hills for the construction of its 200 MW data centre, several problems arose. The first was the electricity availability requested, which was too much for the electricity supplier, whose energy mainly came from coal plants. The construction of a new plant of this type to supply the data centre would have required major financial investments and would have adversely impacted the installation's energy balance. Black Hills and Microsoft therefore made a deal: the data centre would be equipped with gas backup generators that could be put into operation to relieve peaks on the local electricity grid by returning electricity to it during peak hours. The data centre consequently became a cutting-edge power plant, based on cleaner gas generators. In exchange for this, Black Hills would buy electricity on the market and route it to Microsoft if there were production shortfalls. Microsoft benefitted from a negotiated rate (when Microsoft's load exceeded 35 MW). This rate was intended for large industrial customers who used less than 13 MW of electricity, had a large quantity of backup energy onsite and were ready to ask the electricity company to access this energy source to make up for deficits caused by peaks.

Let us further mention Portland Dispatchable Standby Generators, a smart grid by Portland General Electric (PGE), the principle of which is to remotely mobilize backup generators (using diesel oil). Thirty-five customers thus made available their 86 generators, including data centres (six of Via West, two of TATA and one of ODAS), as well as hospitals, factories, universities, banks and food warehouses. The networking of these generators made it possible to have a reserve of 121 MW that could be mobilized by PGE to ensure part of these mandatory emergency resources (Figure 11.5).

There is a pooling between different actors and a 'win-win' situation because PGE supplies fuel oil and handles maintenance of the generators

Figure 11.5: Portland Dispatchable Standby Generators, photos of backup generators at the Via West data centres and other buildings connected to the micro-grid

Source: © Fanny Lopez, 2017

for 15 years. Moreover, during monthly tests, the energy produced is recovered by the grid. In the different examples presented, the idea is to pool the dormant emergency capacities by connecting them to each other or sharing them to lighten the load on the electricity grid. The problem remains, however (for certain examples), in the use of fuel oil as the primary energy source, which reduces the virtuous impact of the pooling approach in terms of the carbon footprint.

Despite calls for multiplying this type of exchange, examples remain rare. In Sweden,[5] but also in the Netherlands, the idea of buying and selling the electricity of backup systems is being increasingly discussed as much by the electricity operators as by the data centres. Apart from electricity, it is the request for heat that seems to attract certain public energy operators

Laborious heat recovery

In the United States, as in Europe, the difficult recovery of heat bears witness to this phenomenon. In France, in 2012, in preparatory work for the regional climate, air and energy plan (SRCAE), the Regional and Interdepartmental Directorate of Environment and Energy identified the major challenges

5 Certain agreements with existing customers already in the Stockholm region. On 10 October 2020, the inspector of the energy market presented the results of a large-scale survey to the government, which led to legislative changes. To do so, experimentation is encouraged. Consequently, the electricity operators are carrying out tests, in conjunction with the industrialists, to determine the limits of the current legislation and possible future adjustments.

linked to heat recovery from the data centres, whose average temperature was about 40–50°C, which was considered low temperature (<60°C). The local energy agency of Plaine Commune (ALEC) also formulated a recommendation to go further in depth on this subject, which was partially concretized with an ADEME study on waste heat.[6] In the latter study, the authors estimated that the total maximum reserve of waste heat in data centres in Île-de-France amounted to 490 GWh (concentrated in Aubervilliers) out of the 26,500 GWh analyzed.[7] But this estimate is to be treated with caution because it was conducted only from information transmitted by the data centre operators that wanted to respond to the questionnaire.

The integration of heat into the urban heating networks is of interest to managers of the CPCU (Compagnie Parisienne de Chaleur Urbaine) because this does not necessitate capital investment expenditure. However, implementation is difficult. Among the technical curbs, the study underlines the 'non-durability of the resource', that is, that the increase in performance could cause a decrease in the waste heat emitted. Two parameters seem to block this and explain the standstill in terms of waste heat recovery.

The first parameter is the economic barrier: economic profitability models and divergent interests. For heating networks, contracts restrict commitments to durations of 25, 30, even 50 years, whereas data centres aim at returns on investments in two to five years (for customers that are often outside the territory potentially served by the heating network being studied). The interests and timeframes of the projects differ.

The second parameter is the technical barrier: it is preferable to plan for heat recovery when the data centre is being built because intervention and work on the existing building can disturb its operation. The cost of existing systems to be changed and technical readjustments can also be prohibitive. The opportunity for a new construction or renovation seems more relevant, even if the connection distance and necessary temperature must be closely studied (the heating networks in Île-de-France generally have temperatures between 60 and 110°C). For the connection to new buildings, the relevance of such a connection and the related costs must also be evaluated because with heat regulation standards, temperature needs are not as great.

Many of the studies conducted had no follow-up. In France, an opportunity study was carried out in Courbevoie in 2016 in the framework of the Village Delage urban project in order to examine the possibility of putting the

6 https://www.enrchoix.idf.ademe.fr/ressources/chaleur-fatale/etude-potentiel-production-valorisation-chaleur-fatale-idf.pdf [consulted in December 2018].

7 ADEME study, p 23. 'The maximum reserve of waste heat in Île-de-France, for the perimeters of wastewater, industrial waste heat, waste incineration units and data centres, is estimated at about 26,500 GWh, or 22 per cent of the demand in fuel from the residential, service and industrial sectors in the Paris region.'

existing heat sources to good use. This study only focused on the data centre that was going to be relocated during the restructuring of the district, that of Telecity. As the future of the two other data centres (Telco-DCIM and SFR, which had at the time 10,000 m^2 of hosting facilities) was not known at the time, heat recovery was not studied. The conclusion of this study was 'that it is almost impossible to anticipate what a data center will be in 10 years' and that heat recovery could not be envisaged for Courbevoie, which has however, an urban heating network. In Clichy-la-Garenne, in 2007, feasibility and technical studies, which were not followed up, were conducted on the 50,000 m^2 data centre of Global Switch, the former BIC lighter factory transformed by Reid Brewin Architectes.

The Idex data centre, which manages the heating network of the cities of Clichy and Levallois, did not succeed in heating contract agreements, whereas the technical studies showed that the right technical and economic conditions were present. The architect of the Interxion data centre in La Courneuve, on the rue du Rateau, mentioned the existence of feasibility studies carried out when the building was designed. They also had no follow-up.

Despite certain failures, a few projects were successful. They concerned very local loops, often connected to a building located in the immediate proximity of the data centre. This was the case of the very large-scale urban renewal operation steered by the L'AUC: 'Chapelle International', a new 104,000 m^2 district with a 45,000 m^2 logistics building, including a city of Paris data centre (Figure 11.6). The district has the first heating loop of heat recovery from a data centre. This 500-metre-long loop operates as a closed circuit based on two 270 kW heat pumps connected to the cooling circuit of the computer bays and a backup gas boiler. The circuit is managed by CPCU and covers 50 per cent of the district's heating and hot water production needs.

In Bailly-Romainvilliers in Marne-la-Vallée, in the Le Prieuré mixed development zone, the heat recovered from the 10,000 m^2 of the Natixis bank data centre makes it possible to supply hot water to an aquatic centre and an 1,800 m^2 business incubator. This is also the case of Céleste (Figure 11.7), an Internet service provider and the operator of a national optical fibre network, whose data centre was built in 2012 by the architecture agency Enia in Champs-sur-Marne. The architectural design enables this high-density data centre to be cooled without having to produce cold water. For 95 per cent of the year, it is cooled via the use of ambient air. The heat emitted by the machines is utilized to heat the offices.

In the Beaujon mixed development zone, the Orange data centre heats the apartments above it (an SEM Paris Seine operation). Another example of heat recovery is Iliad, at the former site of the Ponts et Chaussées central laboratory in the 15th arrondissement of Paris. Iliad installed a data centre in a former bomb shelter which had been abandoned since 1991. The heat

Figure 11.6: Data centre of the city of Paris, 2022

Source: © Cécile Diguet, 2022

Figure 11.7: Céleste offices and data centre in Champs-sur-Marne

Source: © Céleste

produced by this ultra-secure data centre, dedicated to archiving, is recovered to heat 150 social housing units (80 per cent of the heat and 50 per cent of the hot water) above the shelter. This partnership with Paris Habitat, of a ten-year duration, makes it possible to recover 250 kW for heating and 700 kW for domestic hot water in the winter. This will represent savings for the inhabitants of €500 a year in building charges. The data centre belonging to the Université de Bourgogne is used to heat the adjacent buildings through the university's heating network. This is made possible because the boiler supplying the university's heating network is located in a building adjoining the data centre. About 5 per cent of the university's heating needs are thus met, approximately 400 kW. Another way of using the heat from computing or storage is to favour separating the elements. This is the principle of the digital heaters made by the Grenoble company Stimergy (liquidated in 2019) and Qarnot heating[8]: the data centre was split into several areas and, by being as close as possible to heating needs, made it possible to economize costs on the network pipes as production is closer from consumption. The swimming pool in Butte-aux-Cailles (Paris, 13th arrondissement) is heated at 30 per cent by six digital boilers (computer servers) by Stimergy. These six boilers made it possible to save on both the heating bill and the carbon footprint, since 50 tonnes of CO_2 are economized each year. These boilers were also installed in buildings or eco-districts in Nantes and Grenoble in social and student housing, with savings of 40 per cent on the heating bill. The servers are placed in an oil bath that absorbs the heat produced and reheats the hot water tank. Thanks to water heated to 45–50°C, iit is possible for the digital boiler to cover 30 to 60 per cent of the building's needs. The Qarnot company's objective is quite similar: it is to move the heat source from the data centre to the residential building. Created in 2010, Qarnot specializes in distributing computing and heat production. The company has rolled out 800 digital radiators, the Q-Rad, in buildings like social housing in Balard (Paris, 15th arrondissement) and the Forestine residence in Bordeaux (in partnership with Gironde Habitat). Qarnot sells its computing capacities to its customers. The computing is done directly in the apartments radiators, enabling them to be heated.

Often on the initiative of one company, these cases, despite their rather isolated nature, bear witness to a search for pooling. There are also public approaches of a larger scope in which land, energy, urban and economic questions are considered: this the case of the Stockholm Data Parks program and Dutch planning.

8 Interview with Erik Ferrand, founder and President of Qarnot Computing, Paris, and Christophe Perron, founder and President of Stimergy, Paris, March 2018.

12

The Dutch and Swedish Examples: Integrated Urban Approaches

Amsterdam was the first European city to take hold of the subject of digital infrastructure spatialization and to propose updated planning. Since 2018, many evaluation reports and documents commissioned by the public authorities have involved all the actors: data centre industrials, electricity transmission and distribution operators, as well as the metropolis' urbanism departments. We can cite the report by the city of Amsterdam, *Vestigingsbeleid data centers, de Amsterdamse regio als Green Data Port*, from 2013, and also especially *Datacenter Spatial Strategy, 2030 Roadmap for the Growth of Datacenters in the Netherlands* (REOS) in 2019. This global strategy is notably based on cooperation at the national (under the direction of the Ministry of the Interior) and regional levels. This report gave rise to a work bringing to the forefront the spatialization of the energy and land impact of data centres, a document unique in Europe.

The REOS[1] presented a new national spatialization of data centres in three illustrated scenarios: (1) metropolitan concentration; (2) regional cluster; and (3) territorial decentralization. The report recommends a two-stage development, first within the metropolitan region (with a mixture of scenarios 1 and 2). It then suggests that the digital infrastructures extend to where maritime wind farms are being developed (on the west coast) and south of Amsterdam (in the Rotterdam/The Hague metropolitan region). It also recommends an expansion of the existing hyperscale sites of Hollands Kroom and Eemshaven. From 2019 to 2022, it has proposed, however, the reduction of the concentration of metropolitan power and

1 *Ruimtelijke Strategie Datacenters Routekaart 2030 voor de groei van datacenters in Nederland*, March 2019.

favours the development of data centres outside the cities of Amsterdam and Haarlemmermeer, for example, in Almere. It also suggests that medium-sized data centres be linked to an urban environment and/or a heating network.

In June 2019, the report *Systeemstudie energie-infrastructuur Noord-Holland 2020–2050*[2] offered an in-depth analysis of the infrastructural and energy inter-relations and the transformations necessary. Whereas the Netherlands is historically a major natural gas producer, the end of gas extraction[3] has been announced by 2030, with the ambition of finding alternatives for heating residential buildings. The control, even the reduction of consumption based on fossil fuels and the increase in renewable energies in the energy mix, are a challenge for the country. Aiming at carbon neutrality by 2050, the Netherlands has been developing, for several years, a policy that incentivizes renewable energies, in particular wind energy and photovoltaic solar energy, which is creating an overload on the network due to rapid requests for additional electricity. With growing environmental concerns, the question of the value added by data centres to the local economy is being raised. Taking into account the ambitions of the city of Amsterdam in terms of sustainability,[4] their development is, today, being questioned, all the more so as the pressure on urban substations is providing a glimpse of conflicts of uses.

The report *Thematische studie elektriciteit Amsterdam* by the urbanism department of the city of Amsterdam and the electricity distribution operator Liander (2019) concentrates on the substation situation. In the dense zones of Amsterdam and Haarlemmermeer, any surplus power has mostly been reserved by data centres. In Haarlemmermeer, their electricity demands have gradually increased, going as far as blocking other urban developments.[5] The new Hyde Park neighbourhood, planned for around 4,000 new housing units, is in a shortage situation, waiting for a capacity that was reserved by the data centres.[6] The extension of the grid and the construction of new substations are being studied, but the energy control policy calls for better sharing of the electricity resource between upcoming

2 CE Delft, ECN.TNO and SMV, *Systeemstudie energie-infrastructuur Noord-Holland 2020–2050*, report, June 2019.

3 In 2018, 89.9 per cent of the country's primary energy supply came from fossil fuels (in first place, natural gas, then oil and coal) and 8 per cent from renewable energies.

4 The city is aiming at zero CO_2 emissions and the end of natural gas consumption by 2040.

5 Liander was led to conclude special agreements with the *data centres*, forcing them to no longer supply standard redundancy on the grid because there was no longer any available power, while waiting for new substations and additional capacities to be built.

6 Interview with Mariëtte Sedee, Burgomaster, regional and environmental development, agricultural affairs, city of Haarlemmermeer; Jeroen Graafland, Head of the Urbanism Department, city of Haarlemmermeer; Clément Torre, urbanist, city of Haarlemmermeer; Irma Graveland, city of Haarlemmermeer, March 2020.

urban developments and digital needs. By crossing the data of the TenneT transmission operator and the Liander distribution operator[7] with urban projections, various scenarios have been forecast.[8] They make it possible to identify between today and 2050 the impacts on the electricity grid of urban developments like the construction of new housing, the deployment of electric cars and the construction of data centres. A saturation of substations is anticipated by 2030. In 2050, the electricity demand will be two-and-a-half (low scenario) to five times (high scenario) higher than it was in 2018.

As can be seen on the diagram in Figure 12.1, the data centres (in light green) have the greatest impact on the electricity grid. The load is about 750 MW. This power increases to 2,000 MW (in the low scenario) and 4,000 MW (in the high scenario) in 2050. In the medium scenario, the load will be about 2,900 MW in 2050. The data centres and industry have the greatest impact, followed by new constructions and mobility.

The map in Figure 12.2 presents the location of the existing substations, with the available energy capacities for each. The small green dots indicate sufficient capacity, the orange dots indicate that the capacity limit is almost reached, and the red dots indicate insufficient capacity. The first line is for 2030, below 2050. Each scenario – low, medium, high – is represented in the grey boxes.

In Amsterdam, in 2019, there were 25 substations (Figure 12.3). In 2030, 11 (in the low scenario), 17 (in the medium scenario) and 23 (in the high scenario) will run into their capacity limits if no measures are taken. The document points out that in 2030 (medium scenario), 37 per cent of the total electricity demand will come from data centres. Their impact is concentrated on three zones of the city (scientific park, Amstel III and the port zone). To solve the bottlenecks in the substations, the grid must be expanded by adding, before 2030, six to eight substations (low to medium/high scenario) on new sites (2,000–5,000 m^2) in Amsterdam.

---

7 Interview with Ruud Berndsen, Head of European Affairs, Alliander; Sander Oosterloo, Network Development Expert, Alliander; Roelof Potters, Head of Energy Transition, Alliander; Bert Ross, Head of Relations and Environmental Strategy, Liander; Ter Braak, Regional Development Expert, TenneT; Robert Kuik, Head of Grid Development and Strategic Assets Management, TenneT; Ruban Rasing, Head of Grid Development, TenneT; Chantal Braak, Head of Grid Development, TenneT; Peter Kwakman, Head of Grid Development and Data Centres, TenneT; Joost Molenaar, urbanism, public services strategy, utility strategy team, Amsterdam; Petra Boegheim, urbanist and Head of Data Centre Strategy, Amsterdam; Ruben Voerman, urbanist, regional development and sustainability, Head of Municipal Electricity Strategy, Amsterdam; Tim Ruis, urbanist, regional development and sustainability, Head of Municipal Strategy for Urban Heating, Amsterdam, all in March 2020.

8 City of Amsterdam and Liander, *Thematische studie elektriciteit Amsterdam*, March 2019.

Figure 12.1: Low, medium and high scenarios in MW

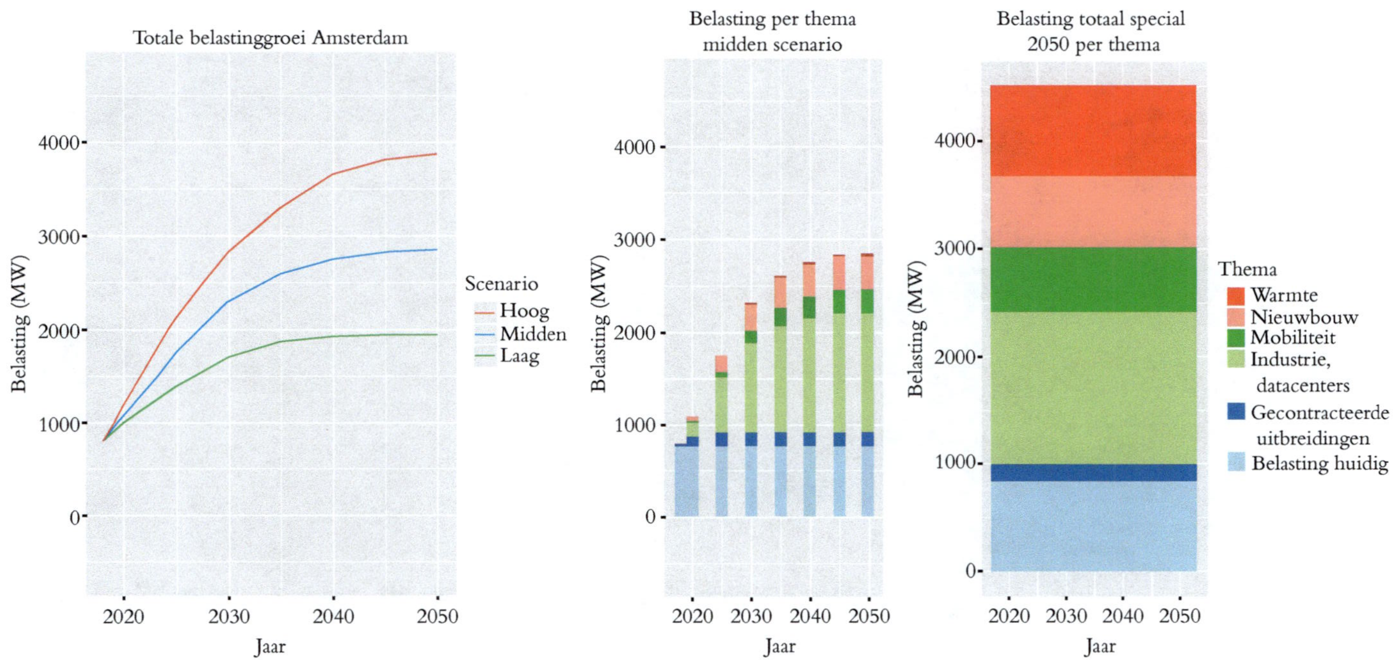

Source: Thematische studie elektriciteit, Amsterdam, 2019

Figure 12.2: Electricity availabilities in the substations

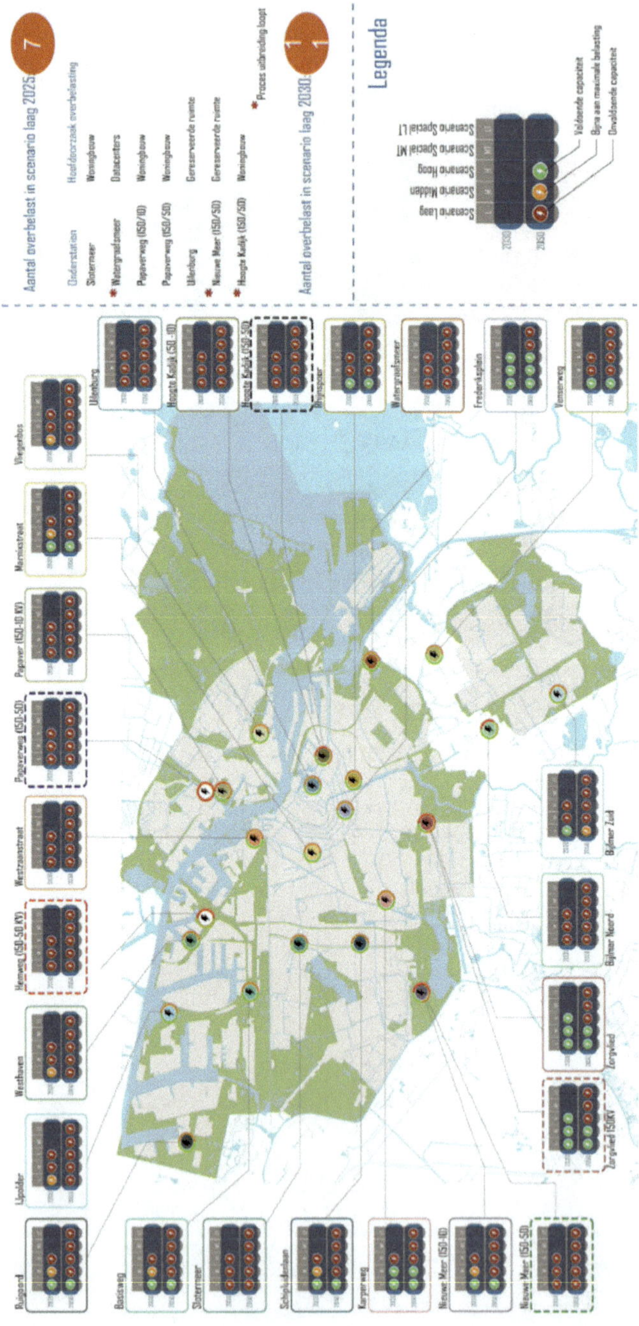

Source: City of Amsterdam and Liander, Thematische studie elektriciteit Amsterdam, March 2019

Figure 12.3: 'Substation' buildings

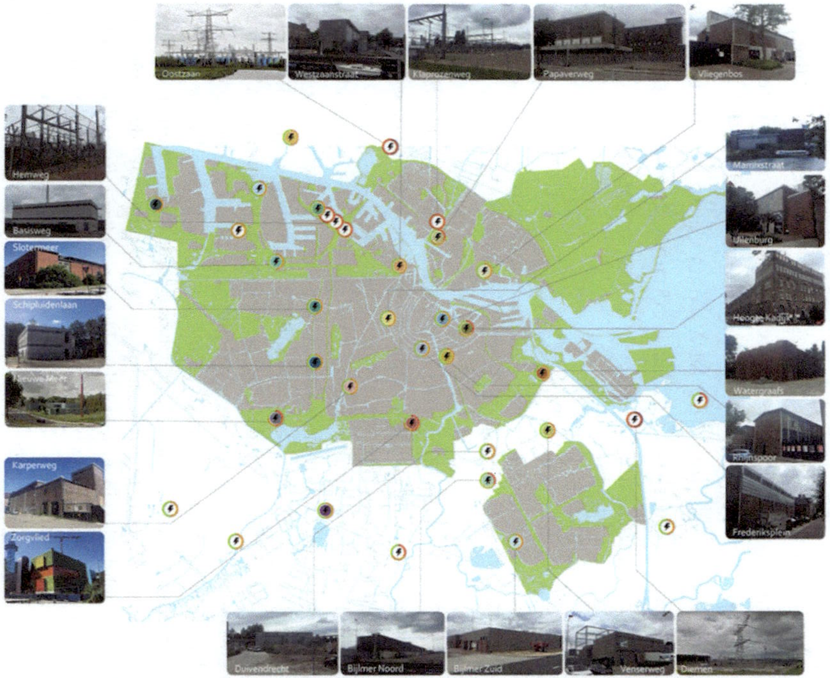

Erontbreken 5 afbeeldingen: IJopolder, Ruigoord, Weshaven, Tweeschakelstations (Zuidas)

Onderstation Zorgvlied (Zuidas)

Onderstation Watergraafsmeer (Sciencepark)

Source: City of Amsterdam and Liander, Thematische studie elektriciteit Amsterdam, March 2019

The city of Amsterdam chose the strong impact scenario to find space for new substations. Not all of them are going to be built immediately, but land reservations have begun. The city decided to only build in the already urbanized zone. There will be no extension on farmland. One of the results of these two years of reflection is the securing of a digital hub of 150 hectares south of Schiphol for an exclusive data centre function with an electricity consumption of 1 GW, half of that of Ashburn. Concurrently with the digital reflection, a national reflection was carried out on the location of regional pockets of 1 GW. TenneT, Liander and the national government mobilized around these 'electricity nodes'.[9] Potential sites are examined vis-à-vis connectivity, electricity and spatial availability and the possibility of using the data centres' residual heat for future residential developments.[10] It is estimated that the preparation of such a cluster would take at least five to ten years to develop. Radical measures have been taken to accompany urban development. The creation of a 1 GW load pocket represents a substantial investment that shows the determination to support the growth of the digital sector, a synonym for the public policy on the region's economic development.

In the Amsterdam metropolitan region,[11] the new regulations from the moratorium include obligations concerning energy and water consumption for cooling, but also for reusing residual heat[12] and spatial integration. For example, large-capacity (over 80 MW) data centres must install their own substation to avoid overloading the electricity grid and to guarantee that enough electricity remains on the grid for other needs. They must buy

9 D'Cision, *Kansen en bedreigingen voor datacenters in de Metropool Regio Amsterdam*, 2019; and Buck Consultants International, *MRA-brede Strategie Data centers*, CE Delft, 2020.

10 Uncertainties on the sizing of the new cluster in terms of size and capacity persist. For certain public actors, the growth scenarios estimated by the Data Center Association lobby seems very ambitious, just like the estimates concerning their energy consumption.

11 Interview with Joost Molenaar, urbanism, public services strategy, utility strategy team, Amsterdam; Petra Boegheim, urbanist and Head of Data Centre Strategy, Amsterdam; Ruben Voerman, urbanist, regional development and sustainability, Head of Municipal Electricity Strategy, Amsterdam; Tim Ruis, urbanist, regional development and sustainability, Head of Municipal Urban Heating Strategy, Amsterdam, all in March 2020.

12 The recovery of residual heat near residential zones is encouraged; however, in the Netherlands, the heating networks are not very developed and serve less than 4 per cent of housing units. New developments are planned, but the question of temperature remains problematic. In 2019, the regional planning and sustainability department of the city of Amsterdam published a document titled *Het Amsterdamse Bronneboek*, which provides a glimpse of the different sources of waste heat produced in the the city, taking into account the different temperature levels. The challenge consists in creating links between heat supply and demand, notably concerning lower-temperature heat. To this day, there are still major uncertainties that block the integration of data centres into the urban heating networks. Financial models, regulatory questions and large needs in electricity for high-temperature heat pumps remain obstacles.

renewable electricity and make their residual heat available free of charge so that it can be used for urban heating. The contribution of data centres in terms of onsite electricity production, storage or heat recovery is increasingly fostered without being mandatory at this stage, unlike in the Swedish capital.

The Data Centre Parks of Stockholm

Stockholm has developed a program that forces data centres to be integrated into the urban heating network and therefore to be installed on the network's specific sites. Initiated in 2014, in the framework of a partnership created between Stockholm Exergi (a municipal subsidiary of the electricity company Fortum[13]), the city of Stockholm, the electricity operator Ellevio and that of dark fibre, Stokab,[14] the Data Centre Parks program represents a major initiative in terms of planning and energy control. The interest is to rationalize and control the urban siting of data centres, while favouring territorial energy efficiency. The Stockholm Exergi company was created in 2018 and is held equally by the city and Fortum, the latter steering this project. For 2030, the ambition of Stockholm Exergi is to eliminate 10 per cent of fossil fuels from its energy mix and to replace them through heat recovery.[15] One of the specificities of the city of Stockholm is having a very efficient heating network, whose 2,800 km are supplied by five large heat and electricity production units.[16] With demographic growth, the objective is to ensure heating needs[17] while reducing CO_2 emissions.

13 Fortum is one of the main Nordic energy producers.

14 Stokab is a public dark fibre company, an exclusive provider of infrastructure and not of services. It has belonged to the city of Stockholm since its founding in 1994. Its creation resulted from the city's ambition to invest in long-term open and neutral digital infrastructure planning to guarantee fairer competition in service provision. Stokab has a very strong discourse on the idea of public service as the basis of the development of services and innovation (notably for SMEs), thus contributing to growth and job creation. The deployment of the network is limited to the Greater Stockholm region and the coverage ensures the multifibre FTTH connections and covers 90 per cent of households and almost all commercial buildings. It is one of the largest dark fibre networks in the world.

15 Interview with Erik Rylander, Head of the Stockholm Data Parks program, Stockholm, October 2018.

16 Of a total capacity of 4,000 MW of heat and 650 MW of electricity.

17 The success of heat recovery is due to the fact that the Stockholm network is the largest heating network of the Nordic countries (and the world's largest cooling network). The coexistence of heating and cooling networks allows for flexible and economical solutions. In this cold climate, heat has a value and the country has over 25 years of experience in excess heat capture and reuse. We can also mention the heat recovery of the interconnection centres of the Stokab network, which is undertaken in the Östra Reals school. The heat emitted by the connection is converted in the school's heating system, which makes it possible to reduce electricity consumption by 30 per cent.

Sweden is very advanced in the development of renewable energies and recovery, and it is from this perspective that Stockholm Exergi started to reflect in 2012 on a program to recover heat from data centres. 'A data center load of 10 MW can heat around 20,000 residential apartments', points out Erik Rylander, head of the energy project. The Stockholm Data Parks program decided, in 2018, to specialize four sites to house data centres (of over 5 MW) in order to control their expansion. In exchange for this location, the municipality offers them: renewable energy at negotiated rates; connection of the building to the fibre network; rapid and accompanied measures (authorizations); and free use of the city's cold network when the data centre has a load of more than 10 MW. As for the data centre, it contributes to the local energy system through the recovery of its heat, which is pooled and made available. An option permits the data centre to manage its cooling itself. Stockholm Exergi then buys the surplus heat, at a price reflecting its production costs.

The Stockholm Data Parks program initially comprised four parks: Kista (open in 2017), Brista (open in 2019), Skarpnäck and Väsby (whose opening was planned for 2021, but is still pending for Skarpnäck and abandoned for Väsby). All are located on municipal land, leased to the operators in exchange for the recovery of their residual heat. The Kista district (Figure 12.4) is a service sector many of whose services are dedicated to the digital sector. It is in immediate proximity to residential and university zones (Stockholm University and the Royal Institute of Technology). Part of the Kista district,

Figure 12.4: Former printing plant transformed into data centres in Kista

Source: © Fanny Lopez, 2018

Figure 12.5: Stack of the Fortum plant, urban heating system in Kista

Source: © Fanny Lopez, 2018

near the highway, is not specialized in housing data centres, and car dealers can mobilize their parking lots over the longer term to densify the zone. Interxion has six data centres there on two sites, one of which is a former printing plant that was recently transformed and connected to the heating network. Nearby, Fortum has a heat and cold production plant (Figure 12.5) and will soon have as a neighbour a new data centre of the AtNorth company that was due to complete, in January 2022, the construction of its first 12 MW model, connected to the heating network. Two other projects are being studied in Kista, but remain confidential at this stage.

The second data park, Brista, is near the airport and extends over an unbuilt space between a railroad line and an electricity and recycling installation. In 2018, 200,000 m^2 of land and 30 MW of power were available on this site.[18] According to a 2019 press article,[19] it is: '20 MW of power available today, 30 MW in 2022, with the possibility of an extension to 100 MW in

18 Erik Rylander, Stockholm Data Parks, 'Stockholm – a circular city: The role of the data center as a resource in city infrastructure', April 2020, Available from: https://www.datacenterdynamics.com/en/conferences/energy-smart/2020-virtual/stockholm-a-circular-city-the-role-of-the-data-center-as-a-resource-in-city-infrastructure/#presentation [Accessed 25 February 2025].

19 Energy Smart: Stockholm Data Parks announces 30MW campus, DCD, Available from: https://www.datacenterdynamics.com/en/news/dcdenergy-smart-stockholm-data-parks-announces-30mw-campus/ [Accessed 12 December 2020].

2030. Additional land of 400,000 m^2 could also be added.'[20] In 2022, Erik Rylander, head of the program, said that the Brista park will be developed by a major industrial actor (confidential here too) that will restore its waste heat, but is not a data centre.[21] The Data Parks program planned two other sites, in the Stockholm area, in Skarpnäck and Väsby, both of which had a heat recovery system. Skarpnäck has still not been officially launched due to uncertainties concerning the site's electricity supply. Väsby was withdrawn from the program because of a lack of electricity grid capacity and limits on the distribution potential of the recovered heat. Despite integration efforts, tension has remained high. Other enterprise or colocation data centres are benefitting from the heat recovery program without being in the data parks: Bahnhof (three contracts), Ericsson (two), H&M (two), ComHem, Bankgirot and GoGreenHost. For Erik Rylander, this project's success factors are about all the robustness and size of the urban heating network; the number of customers for the heat (need for 12 TWh/year); a clear technical interface; the transparency of the system; and the predictability of heating needs. This strategy has made it possible to channel the development of data centres. The city clearly planned ahead: it attempts to find sufficient incentives to orient their installations to existing sites and existing buildings or those to be built. It targeted sites having good electricity and connectivity capacities in 2017, but the situation seems to have changed for Skarpnäck and Väsby in 2022. Erik Rylander now considers that the current restrictions for future expansion are linked to the electricity grid capacity and the availability of land near the urban heating network. The problem of congestion on the national network should be solved by investments enabling Skarpnäck to go from 10 to 40 MW of available capacity.[22] He also stresses that the program's success obviously depends on the analysis of the profitability of heat recovery, which will depend on the price of electricity and the alternative cost of the heating. In his opinion, to lower the entry threshold in heat recovery, it is important to develop servers that function with a higher cooling temperature, as well as low-temperature urban heating systems. Thanks to these solutions (which are already available today), heat recovery and reuse with no energy penalty (that is, without a heat pump) can be applied, which considerably reduces investment and operating expenses.

[20] Interview with Erik Rylander, Head of the Stockholm Data Park Program, October 2018.

[21] Interview with Erik Rylander, Head of the Stockholm Data Park Program, January 2022.

[22] Interview with Erik Rylander, Head of the Stockholm Data Park Program, January 2022; interviews with Ulf Moberg, Technical Director, Svenska Kraftnät; Catarina Grenemark, Head of Customer Relations at the regional grid, Johan Sjöborg, Network Analyst, Hans Larsson, Electricity Systems Analyst, Vattenfall Eldistribution AB; Stefan Lindbom, Commercial Projects Manager for customer relations; and Henrik Bergström, Head of Public Affairs, Ellevio AB, all in September 2020.

One of the particularities of the Kista site is the transformation and reuse of wasteland. Where it was possible, the city paired the urgency for heat recovery with an objective of densifying already urbanized fabrics and land optimization. The digital activity zone model has been strengthened here, but, faced with a very rapid and opportunistic development of data centres, the approach was considered pragmatic. Except for these very controlled developments, the arrival of new data centres in the metropolitan area has been suspended. The Stockholm region, which is the country's largest urban conurbation, is also the one that consumes the most electricity, whereas production there is not very high. In a radius of 200 km around Stockholm, tension on the electricity grid could restrict many urban developments underway or to come. The situation is particularly difficult in winter. This problem started about ten years ago and is worsening. The distribution operator Ellevio is worried about not being able to meet everyone's needs – for example, electrification of the transportation system, enlargement of the metro network, construction of housing. But apart from the question of electricity availability, there is also a question of space. Örjann Lönngren, Policy Analyst for energy and climate at the city of Stockholm, asks: 'How can new infrastructures be planned in a dense city? How can enough electricity be economized to make everything that we want to have function on our territory at this moment? What do we really need? Things must be rethought in-depth.'[23] The arrival of data centres goes beyond the economic, electricity and urban issues of just the territory concerned. The phenomenon has always spilled over the city's boundaries, notably because the balance of the electricity grids is thought of on the national scale. Many experiences have shown the necessity of having not only a broader and coordinated view between the different actors, but also a spatialized one.

In the United States, as in Europe, in the countries the most affected by the development of data centres, the difficulties encountered by the transmission and distribution electricity grid operators are similar and the solutions essentially come from a development vision. The definition of location criteria appears as a key point of rigorous public policy concerned about an economy of means linked to location and the proper sharing of electricity loads and land dedicated to this industry. Depending on the size and type of the data centre, installation on an unused, already urbanized, large plot of land (wasteland) is encouraged, in the same way as the integration into or proximity to an industrial zone or an infrastructure, or near the urban heating network for smaller-scale developments. It is also usually recommended that the ground floor or a part of the building be shared with

[23] Interview with Örjan Lönngren, Policy Analyst for Energy and Climate at the city of Stockholm, October 2018.

other functions, such as stores, services, offices and/or restaurants. The scale and form of the infrastructures, the landscape integration and the mix of uses also seem to be at the heart of controlling digital infrastructures and they involve in the same way public entities, whether they are local authorities or grid operators. Despite the involvement and coordination of the actors that have succeeded in taking control of the phenomenon through a spatial and energy strategy (the Amsterdam and Haarlemmermeer moratorium, the Stockholm Data Parks program), the structural constraints remain strong: energy production and distribution has limits (in terms of resources), timeframes for strengthening the grids have time-based regulations, etc. Many countries are having trouble modifying the procedures and making the historic regulations evolve (prioritizing uses and envisaging energy quotas, for example). Working with the industry, investigations are underway in order to test the relevance and possible feasibility of certain systems, before a change in the legislation. With the digital, the spatiality of the electricity grid has returned to the forefront, notably with the multiplication of substations and the possibility of better architectural integration of these infrastructures in dense zones. Experimentations of local micro-productions integrated into micro-grids are also the object of reflection. The use of backup generators at heavy load hours seems to be possible and is already being implemented or tested. But it is also more largely the questioning of certain taboos like the historic rule of 'first come, first served' that are increasingly being debated faced with the development of a private-sector industry that lacks a framework and against which the public service of energy and urbanism remains perhaps the best protection. Everywhere in Europe, the necessity of a national debate on digital needs is emerging to attempt to control a rapidly expanding infrastructural development whose financial means are inversely proportional to public service means and energy and land resources that are increasingly limited by the environmental crisis.

As for any network infrastructure, alternative infrastructures, often managed by associations or cooperatives, by citizens and enthusiasts, which aim at broadening access as well as encouraging a technical reappropriation of the infrastructure, even its transformation in terms of scale and functioning, correspond to the institutional, large-scale and commercial version. The history of these alternatives often presents conceptual and experimental antecedents and they are part of the search for a free and open Internet. They are based on citizen reappropriations, which partially aim at changing the centralizing and monopolistic scale of the network largely in the hands of private economic actors heavily dependent on Big Tech.[24]

[24] See Félix Tréguer, *L'utopie déchue. Une contre-histoire d'Internet 15ème–21ème siécle*, Paris, Fayard, 2019.

PART IV

Another Digital: Rethinking the Network and Its Infrastructures

The projects that follow, aiming at making the digital technical system evolve towards more diversified forms favouring a control of and decrease in electricity consumptions, resilience to extreme climate events, small-scale network infrastructures and more collaborative functioning. A part of these initiatives has been rolled out through distributed and peer-to-peer modalities, transforming the Internet's technical system at its core, while others aim at taking control over the infrastructures of the existing system to self-manage on more local scales, or, notably for the French state, to take control for geopolitical reasons.

Beyond the question of storage, turning to the network infrastructure is necessary to grasp all the advantages, motivations and approaches of the Internet's associative and cooperative actors. It is entirely in the interest of citizen Internet Service Providers (ISPs) that aim to provide a connection not only to the Internet in remote zones or those abandoned by the operators (citizen Wi-Fi networks, mesh networks), but also to a neutral, unfiltered, unsorted Internet through, for example, a Virtual Private Network (an encrypted tunnel), a system widely used, notably for political reasons in China, Iran and Egypt, for example. The entire structure of the infrastructural system is thus rethought. On this basis, the question of alternative storage (distributed on user terminals or in nearby data centres) comprises a second stage that these ISPs wish to attack, as shown by the Spanish example of Guifi.net. It would therefore seem that the citizen reappropriation of data storage goes hand in hand with that of the network. Whereas computer terminals are more and more locked by constructers, Internet users having very variable digital literacy,[1] and peer-to-peer tools having trouble in

[1] The concept of digital literacy designates a technical and cognitive skill paired with a critical attitude towards the available information.

developing on this context, associative and cooperative movements aim at offering a local and free version of a worldwide Internet, favouring the geographic inclusion of remote territories. At the same time, the question of the states' digital sovereignty is a recurring political subject, for which the geographic location of the storage infrastructures represents an important issue, but whose control is difficult. For all these alternatives, the ecological question remains just below the surface in the discourses, even though their frugality places them in a transformation framework.

13

Cyberattacks and Political Intrusions: States Seeking Independence

States' sovereignty is increasingly based on their capacity to control digital interference, protect themselves from the pirating of public services and businesses, and fight against the theft of strategic data. China and Russia have created national and closed Internets with this in mind. The creation and support of heavyweight digital actors like Alibaba, Tencent and Xiaomi in China, and Amazon and Meta in the United States, also serve these objectives. In Europe, most countries, including France, have repeatedly raised the question of their digital sovereignty.

The subject of secure data storage is at the heart of this concern. A new strategy was consequently adopted in France in 2021 through the label 'Cloud de confiance' and the 'Cloud au centre' strategy. French actors today use Cloud services while making use of a hosting offer of the data centre type. The two options are activated given the variety of needs, notably of the large local communities or public structures, and both of them can be public or private. The rules and offers for the government administrations are different from those of local communities. For both types, the use of Cloud solutions or data centres operated by actors outside Europe is, however, increasingly raising questions surrounding the location and sovereignty of the data. The protection of public data is a strategic issue and foreign intrusions in national affairs, like Russia in the 2016 US elections, has reactivated this issue in France. Consequently, the French state has, for example, prohibited (with exceptions) the use of Microsoft 365 by the ministries in a circular published in September 2021.

As for the local communities, the General Division of Enterprises (linked to the Ministry of the Economy, Industry and the Digital), the General Commission on the Equality of the Territories (linked to the Ministry of Territorial Cohesion) and the Caisse des Dépôts et Consignations contributed to the publication in 2015 of the *Guide sur le cloud computing et les datacenters à l'attention des collectivités locales*. This guide presented Cloud computing and infrastructure pooling solutions that can exist for local communities. What is put forward is that the so-called data centre 'of proximity' projects, projects that pool infrastructures, services and skills between communities and different territorial levels, could favour access to hosting services for smaller-sized communities that do not necessarily have computer and logistics resources. In the same way as for the examples of higher education and research, this solution could make it possible to meet the needs of communities and public actors, while enabling economizing energy in these infrastructures. Nonetheless, few projects of this type have emerged since 2015. The university world in Nouvelle-Aquitaine[1] started to implement a pooling project in the second half of 2010. In 2021, a territorial digital coordination agreement was signed to strengthen a well-thought-out and collective development of the digital, notably by initiating a pooled data centre project. This was between the city of Paris and the AP-HP (the public hospital system), located at the Porte de la Chapelle in Paris, comprises a rare example of this approach.

Other than the question of the building, that of the Cloud tool and its sovereign character is a recurring subject for the French state. It has supported several French Cloud initiatives that failed in the past, like Cloudwatt and Numergy. In 2021, the 'Cloud au centre' strategy and its 'Cloud de confiance' label, aimed at certifying certain operators, were created. The latter incorporates an already existing label, SecNumCloud from the National Agency of Information Systems Security (ANSSI), which guarantees high security on a technical level. Legal elements have been added to block the extraterritoriality of certain foreign laws like the Cloud Act and the Foreign Intelligence Surveillance Act, for example. The 'Cloud au centre' policy is a new doctrine that requires the state's digital services to be hosted either in an internal Cloud or an external Cloud, but one labelled 'Cloud of confiance' by the ANSSI. For the internal Cloud, there will be no more than two inter-ministerial offers: by the Ministry of the Interior and the Division of Public Finances. The debate is continuing in the professional milieu of the digital and the administration. Must the entire Cloud production chain (energy production, design and production of material and micro-processors, storage systems and even operating and virtualization systems) be controlled or not be genuinely sovereign? Would a French or European Cloud really have the capacity to develop, faced with the US heavyweights? Do the states have the means of adequate control to ensure trust in this new label?

[1] Sciences Po Bordeaux, Bordeaux Sciences Agro, Université de Bordeaux Montaigne, Bordeaux INP, Université de Pau et des Pays de l'Adour, La Rochelle Université.

Lastly, a question that is not often discussed is that of the environmental impact. Could this label of trust be associated with environmental criteria in the future to combine digital sovereignty and a limited ecological impact? The quest for sovereignty as it is approached by the French state does not fundamentally call into question the current architecture of the Internet, unlike the initiatives to follow. It seeks, through regulations, to encourage new sovereign Cloud offers that can be labelled through the establishment of technical specifications that are considered to be protective.

14

The Peer-to-Peer Solution: Organizing a System Distributed among Equals

Peer-to-Peer (P2P) computer technology is reminiscent of the beginnings of the Internet in the 1970s and 1980s, when each computer communicated with the others without any intermediary, censoring or filtering, and without data centres. The P2P principle is based on a network 'of equals' in which each machine or terminal plays both the role of customer and server. It also makes possible the exchange of files through protocols like Bittorrent. The spatiality of this distributed Internet remains elusive and moving, since a peer-to-peer system is composed of a dynamic network of participating nodes that go back and forth, while enabling the network to remain functional. Despite its confidential nature, a movement to rediscover peer-to-peer advantages is underway. This alternative architecture is now being used and tested, for communication networks as well as data storage. It is even the subject of a methodological guide like the fanzine *(Re)building Technology*, a development by the Open Technology Institute and the Detroit Community Technology Project, to help local communities set up these new digital tools.

The mesh networks, communicating peer-to-peer

The mesh networks are most often the basis of citizen Wi-Fi or local communication networks, often accompanied by Commotion software, which permits their creation. Primavera de Filippi and Félix Tréguer thus present the mesh network as 'a group of communicating devices linked to each other in a distributed manner, wireless and without any central hierarchy'.[1] All these terminals, smartphones, computers and tablets

[1] Primavera de Filipi and Félix Tréguer, 'Les réseaux Mesh', in Cécile Méadel and Francesca Musiani (eds), *Abécédaire des architectures distribuées*, Paris, Presses des Mines, Collection I, 2015.

communicate with one another, between peers, via Wi-Fi, using a dynamic routing protocol that makes it possible, when a terminal is turned off, to configure a new route for the data packages . Each person therefore becomes a network relay. The article's authors consider that the mesh networks are a breakthrough technology 'in comparison with the classic wireless centralized solutions with a base station or with mobile phone systems'. It is a distributed communication system that can function without the Internet and makes sharing its users' Internet connection to others possible.

The mesh networks have many advantages. They present 'a tolerance and great robustness in terms of network disconnections and breakdowns in general', they have a low cost (less expensive than satellite telephone systems, for example), and they do not have terminals. In addition, they continuously self-optimize. These networks permit citizens to enter the black box of digital networks. Relatively simple to use, they allow reappropriations. However, there are also limits to these systems: the possibility of losing data, the absence of infrastructures (if there are not enough terminals functioning), transmission errors and security problems. At this stage, these networks cannot replace classic telephone networks (GSM, Global System for Mobile Communications) and their local coverage is often modest, but they are very interesting in zones without infrastructures. Thus, in South Africa and Kenya, the Mesh Potato project was developed with the Village Telco company that proposed, between 2009 and 2016, software and a low-cost box to develop a local telephone system, but it was unable to sustain its economic model.

These citizen networks are also interesting in the event of a natural disaster, as was the case during Hurricane Sandy in New York in 2012, but also in Haiti, Thailand and the Philippines, or as a local development tool (in Detroit since 2012). Moreover, countries that are subject to Internet disconnections or censorship can find an efficient alternative in the mesh network, as in Egypt in 2011, in Catalonia in 2018 and in Iran in late 2019, for example. New applications are being developed combining mesh and blockchain technologies like RightMesh,[2] integrating incentives to make the user's terminal available to the network via tokens, which also make it possible to pay for goods and services. These systems allow everyone to share one or more Internet connections, without all the users (or even a minority) paying for an Internet subscription. They therefore provoke the hostility of commercial Internet Service Providers (ISPs) and classic telephone operators. Moreover, the anonymity of these networks makes states fear abuses related to the darknet, and allows too much freedom of the populations of countries with more authoritarian governments. In

[2] Technology, RightMesh, Available from: https://www.rightmesh.io/technology/how-it-works [Accessed 21 March 2025].

France, Internet connection pooling, and therefore the mesh networks, are largely hindered.[3] On the one hand, the Internet user making her or his connection available must keep traces of those she or he invites. On the other, the HAPOPI 2 law makes it mandatory to secure the connection, subject to a fine or disconnection. Furthermore, the networks must be able to use the Wi-Fi frequencies available, but the possible growth of the Internet of Things, which also functions with mesh technology, could saturate these networks in the future. The authors of the mesh article stress that the government must free up 'white' frequencies (unused radio frequencies) to make room for citizen and associative operators. Among the associative or cooperative ISPs, some propose a connection offer that can be described as the 'last kilometre': they are citizen Wi-Fi providers. In Europe, the most developed of these are Guifi.net in Catalonia (about 38,000 nodes)[4] and Freifunk in Germany. A meshing is also deployed elsewhere in Europe and the world: Wlan Slovenija in Slovenia, Ninux in Italy, Red Hook in Brooklyn, New York, and Buenos Aires Libre in Argentina. Primavera de Filippi and Félix Tréguer, both researchers at the CNRS (National Centre for Scientific Research), describe, in their journal article, what Wireless Community Networks are.[5] Their advantages, however, are primarily technical: flexibility, resilience and autonomy of the network, although they are also social, educational and democratic. They participate in a shared governance of a common – the Internet – contributing the values of transparency, inclusion, the social link, technical learning and an incentive to take part in civic life. Stéphanie Vidal proposed the expression: '(Re)sewing with the threadless', subsequently used by the researcher François Huguet in his thesis.[6] He stresses that these systems favour 'an alternative economic practice that serves a form of urban resilience, a pedagogic practice of community reliance'. Providing Internet access is not only a technical gesture, but it is also a social, educational and economic approach, and digital mediation can also be an infrastructural mediation.

---

[3] Decree of 24 March 2006.

[4] Guifi.net also offers classic connection modes, via asymmetric digital subscriber line (ADSL).

[5] Primavera de Filippi and Félix Tréguer, 'Expanding the Internet commons: the subversive potential of the Wireless Community Network', *Journal of Peer Production*, 6, *Disruption and the Law*, 2015, Available from: https://halshs.archives-ouvertes.fr/halshs-0130663 0v2/document [Accessed 26 February 2025].

[6] François Huguet, *(Re)coudre avec du sans fil. Enquête sur des pratiques de médiation infrastructurelle*, thesis under the direction of Annie Gentès and Jérôme Denis, Paris ENST, defended in June 2016. Notably see the summary of the thesis, Available from: https://medium.com/@francoishuguet/re-coudre-avec-du-sans-fil-ae0d4f28888c [Accessed 26 February 2025].

In the milieu of alternatives for a freer, more sober and decentralized Internet, the initiatives often bring together small groups of people mobilized around systems that struggle (for technical and regulatory reasons) going beyond the circle of initiated activists. Development, expansion or a leap in scale is not obvious in a digital world dominated by the multinationals of the telecoms, of the digital and an order to increase clicks. In this regard, Guifi.net is an exception. This Catalonian foundation lays claim to an infrastructure devised on the principles of commons as defined by the economist Elinor Ostrom[7] and a resolutely different service offering that is constantly growing. Guifi.net is an Internet network created in 2004 to meet unfulfilled Internet needs in this Spanish province. It first started on a rather small scale, but, as of 2007, the members realized that private actors were proposing paying services based on the Guifi.net infrastructure without contributing to the network's sustainability. Guifi.net then decided to create an economic and governance model reconciling service development, job and company creation, services for users and the sustainability of the infrastructure, notably with the invention of a licence (or charter), which all the network's actors had to accept.[8] Today, the network is managed as a common, on the participatory, transparent and open model. Everyone can participate, including private actors, but in the framework of a well-thought-out economic model, with a compensation and equalization system that makes it possible to intervene in different territories, more or less profitable, but also to maintain and sustain the infrastructure. The Guifi.net Foundation was created to guarantee the functioning of this original common model. Four to five service operators are major contributors to Guifi.net, paying €200,000–300,000 per month, and in total, 30 service companies have been incubated by the foundation (with jobs created).[9] The network has several thousand subscribers and volunteers who equally participate in the network's functioning. About €20 million have been collectively invested (via Guifi.

---

[7] Ostrom, E. (1990). Commons are defined by three essential dimensions: the presence of a resource (common good), the nature of this resource not being limited; the management by an open community (commoners) of inhabitants, farmers, neighbours, etc, with a horizontal and inclusive governance, which enforces the rules (access to the community is free and non-discriminatory); and functioning according to transparent rules, chosen by the community in complete autonomy, always covered by the laws of the country concerned. These are access and use rules whose goal is to share and preserve a resource (communing) in order to avoid its over-exploitation and depletion.

[8] Roger Baig, Lluís Dalmau, Ramon Roca, Leandro Navarro, Felix Freitag and Arjuna Sathiaseelan, *Making Community Networks Economically Sustainable: The Guifi.net Experience*, in Proceeding GAIA 16, Proceedings of the 2016 workshop on Global Access to the Internet, 2016, Available from: http://people.ac.upc.edu/leandro/pubs/baig-sigcomm.pdf [Accessed 26 February 2025].

[9] Interview with Roger Baig of the Guifi.net Foundation, Barcelona, November 2018.

net, local communities and users) since 2004 to develop this network. It is now composed of half Wi-Fi (it therefore uses mesh technologies) and half optical fibre, like a classic operator. Guifi.net must, however, very actively fight for the right to develop this infrastructure, faced with the dominant telecom operators and unfavourable European regulations.

Distributed storage, reinventing the peer-to-peer

The functioning of the Internet, before the World Wide Web, and before data centres, was founded on self-hosting and a peer-to-peer functioning, in which the intelligence (technical resources and contents) of the network was found in its ends, that is, in the users' computers. The massive development of the web in the early 2000s led to the development of more centralized technical architectures, creating increasingly powerful intermediaries between the Internet users. Despite this trend, the Gnutella project proposed peer-to-peer storage at the time through specific protocols, fighting against cross-currents.

These distributed storage proposals have been re-emerging for several years faced with the risks carried by the ultra-concentration of data by the Cloud and hyperscale data centres. Certain software programs, Cloud solutions or services, now propose to reuse peer-to-peer functioning or distributed storage on the one hand, or to facilitate self-hosting on the other. Peer-to-peer Cloud solutions have recently made an appearance. As the researcher Francesca Musiani[10] points out: 'The central idea underlying these systems is that the files and contents downloaded by the users in the system are stored, partially or totally, on a storage Cloud composed of part of each user's hard drive, interconnected with a P2P architecture.' Security questions are managed by encryption and fragmentation procedures, as well as duplicate systems to compensate for the users' machines that connect and disconnect. There can be, moreover, a swap of hard drive space against online space as in the case of the Wuala project:[11] 'A user can exchange unused hard drive space in her or his own computer for the storage space on the "user Cloud" provided that the user stays online for a certain amount of time each day.'[12] However, these systems are experimental and raise questions surrounding the confidentiality of the data and the security of the equipment. They are nevertheless part of a movement to rediscover the advantages of peer-to-peer computer technology, which moreover can provide 'answers to questions that concern bandwidth distribution, traffic balance, distribution

[10] Francesca Musiani, 'Stockage distribué', in Méadel and Musiani (eds), *Abécédaire des architectures distribuées*, p 209.

[11] This project came to an end in 2015, having failed to find its economic model.

[12] Musiani, 'Stockage distribué', p 212.

efficiency' on the Internet network more generally.[13] For the researcher Primavera de Filippi,[14] Cloud computing, despite its advantages in terms of flexibility and data availability, raises a certain number of challenges concerning security, private life and the autonomy of Internet users. She thus notes several initiatives that attempt to meet these challenges through decentralized alternatives, while keeping flexibility and availability. She cites several peer-to-peer Cloud projects: the open software SlapOS, OwnCloud and CozyCloud. These projects remain, at this stage, relatively confidential, but constitute interesting potential for users whose mistrust has increased faced with the dominant actors of the digital (Big Tech), in particular, in light of the fraudulent uses of their data (the Cambridge Analytica scandal, among others). Self-hosting services, with remote consultation of one's data, are also developing, through Yunohost, for example. These Clouds therefore favour a decentralized and distributed storage of data and reduce the need for large data centres. They therefore represent the potential, on the one hand, to reduce the spatial impact and land take of these installations and, on the other, to make it possible, like the digital heating offerings of Qarnot Computing (Qalway heaters and radiators), to better distribute storage capacities on the territories. The Catalonian citizen ISP Guifi.net in fact attempted to develop a Cloud common project[15] in the second half of the 2010s. This service, called Cloudy, proposed the same social organization as that of the Guifi.net network (a common), in order to offer a hosting service for third parties. Cloudy represented an edge computing-type service with storage distributed in the Guifi.net users' machines. The simulation of the economic model of a Cloud common led to positive conclusions, but, ultimately, the service did not meet public expectations. Guifi.net still hosts its data in a classic colocation data centre, where one of the Internet's main backbones arrives, in the port zone of Barcelona.

Despite the proliferation of free, open and frugal initiatives, the centralized and dominant digital complex leaves only very little room for a transformation of the global Internet architecture through peer-to-peer solutions. Far from comprising a counter-power on the scale of Big Tech, these proposals nonetheless have the merit of opening the way and making other technical possibilities exist.

---

13 Francesca Musiani and Valérie Schafer, 'Le modèle Internet en question, 1970–2010', *Flux*, 3(85–6), 2011, pp 62–71.

14 Primavera de Filippi, 'Cloud computing', in Méadel and Musiani (eds), *Abcédaire des architectures distribuées*, p 49.

15 Roger Baig, Felix Freitag and Leandro Navarro, Cloudy in guifi.net: establishing and sustaining a community cloud as open commons, ScienceDirect, Available from: https://www.sciencedirect.com/science/article/pii/S0167739X1732856X [Accessed 26 February 2025].

15

Self-Management and Technological Sovereignty: Decentralized Digital Infrastructures

Two approaches coexist, each favouring a different strategy: on the one hand, the first approach starts with distributed protocols and the Internet technical system to transform the landscape of the actors and improve access to the digital (preceding part) and, on the other, the second approach first of all changes he actors' typology: associative and cooperative Internet Service Providers (ISPs), decentralized on the territory. These actors will then help technically modify the digital system by the adaptation to more local scales, and a less centralized organization. Thus, whereas the distributed systems proposed by alternative actors aim to radically transform the Internet architecture, the projects presented here are more involved in a decentralized architecture.

Communicating freely: networks in the hands of civil society

The associative ISPs present many advantages: coverage of territories poorly served by the classic operators, notably rural zones; development of local skills and indirect jobs; accessible rates; involvement of volunteers and users, a social link; and commitment to data protection. Their growth, however, is limited by the number of volunteers available, which makes repairs potentially more complicated in the case of service interruption.

In France, in addition to the four main ISPs (Bouygues, Free, Orange and SFR), there are the associative and cooperative ISPs. French Data Network (FDN) was the first. Created in 1992, it was the first structure to provide Internet access for the general public when the web was first established, before France Telecom in 1996. FDN still exists today and covers the entire

French territory, through asymmetric digital subscriber line (ADSL) and very high-speed digital subscriber line (VDSL) (which offer greater speed).[1] In 2021, FDN opened the first optical fibre line for its members,[2] and proposes protected connection tools through its program Road Warrior Internet Connection for people in countries where Internet access is not certain or secure. Other citizen ISPs propose their services in France: franciliens.net in Île-de-France, Aquilenet in the southwest, Tetaneutral in Toulouse, SCANI in Burgundy and in the Antilles in Saint-Barthélemy, for example, with Igwan.net. These associative ISPs are brought together in the Fédération FFDN.[3] Franciliens.net is an example of an associative ISP in Île-de-France. It proposes an Internet access via ADSL (fibre and Wi-Fi are planned), but also offers an Internet 'brick' to connect to the user's box, combined with a VPN, which makes it possible to access a neutral and open Internet. This brick also enables self-hosting to store the user's own data and create an email account.

As Benjamin Bayart,[4] Co-President of the Fédération des Fournisseur d'Accès à Internet associatifs (FFDN), points out:

> What skills are there on the territories in terms of infrastructure, network maintenance? What is routing and how does it work? How can hosting of a website be done? All that is a cloud of skills around digital questions that develops or not. We find ourselves with a digital territory that is structurally peripheral: there are people who develop websites, people who consult websites, but the sites are always hosted elsewhere, an elsewhere that is very far. The associative ISPs provide economic diversity and therefore a certain wealth. This is what produces resilience. The effect of only relying on just three poor network operators is that infrastructure competence on the network is hyper-centralized in three Paris locations. This isn't all that caricatural, but it isn't that far from the truth.[5]

1 'FDN obviously does not have its own infrastructures everywhere in the country, and therefore rents the services of other operators, called "collection providers," to link your home to the association's servers that are located in Paris. It is then FDN, your ISP, that notably takes care of assigning you an IP address and routing your traffic to the rest of the Internet'; Accès Internet, FDN, Available from: https://www.fdn.fr/services/acces-internet/ [Accessed 25 March 2022].

2 Access to fibre is possible through unbundling.

3 Membres de la Fédération des fournisseurs d'accès Internet associatifs, FFDN, Available from: https://www.ffdn.org/fr/membres [Accessed 26 February 2025].

4 Interview with Benjamin Bayart, Co-President of the Fédération des Fournisseur d'Accès à Internet associatifs (FFDN), 20 September 2018.

5 For more resources on the subject, see the (Re)Building Technology website, Available from: https://communitytechnology.github.io/ [Accessed 26 February 2025]; and The

The SCANI ISP in the Yonne has, for example, enabled connection to the Internet for a fab lab (a fabrication workshop using wood, metal, but also 3D printing, laser cutting, etc), without which it could not have developed its activities: the Grange de Beauvais.

Data centres of proximity and associative/cooperative data centres

Small data centres that could be described as data centres 'of proximity', like a business or a facility, are also one of the options being explored. Among the alternative hosting actors, a collective was created in 2016: Les Chatons as presented on their website is a 'Collective of Alternative, Transparent, Open, Neutral, Solidarity Web Hosts'. It brings together structures wishing to avoid the collection and centralization of personal data in digital silos of the type proposed by GAFAM. This collective is consolidated around a charter that guarantees users the respect of these values. Ecological and energy dimensions are not mentioned in the charter, but this does not prevent projects containing these aspects from being individually initiated. These alternative data hosting structures are found first and foremost in the metropolises: Paris, Bordeaux, Rennes, Nantes and Lyon, and, to a lesser degree, in more rural territories like Exarius in Montluçon and Syntax Informatique in the Orne department, in Domfront-en-Poiraie. Some of them propose locally hosting data, while others partially manage hosting on external data centres like those of OVH, but through free software, and without exploiting the customers' data or using advertising or other tracking services. Les Chatons is a structure undergoing growth, with 92 members in 2023. A recent arrival is the Deux Fleurs association. It recovers old computers that are reconditioned to serve servers, and are installed in the homes of the founders, in Rennes, but also in Orsay in Île-de-France for service provision, and in Suresnes for backup. The spatial organizations are therefore opportunistic, closely linked to the location of the association's members.

Trials are also being carried out for data centres of proximity, more low tech, autonomous and locally managed. This is notably the case for a trial at Rutgers University in New Jersey. These projects, however, although numerous, remain very academic and are not being disseminated outside research circles.

Among the alternative operators having mini-data centres, SCANI is a cooperative community-oriented enterprise that deploys, manages and maintains a communication and Internet access network, which in the longer

community network manual: how to build the Internet yourself, FGV SB, Available from: http://bibliotecadigital.fgv.br/dspace/handle/10438/25696 [Accessed 26 February 2025].

run will offer storage services for the entire Yonne department and which is the property of its users: private individuals, companies and communities.[6] As Bruno Spiquel, one of the active members of the cooperative, stresses, it is relevant to install small local data centres as close as possible to users: 'As hard as it was 10 years ago to find places with a connectivity favourable to hosting a machine, now, between cable, VDSL and fibre, more than 50–100 Mbps of capacity, largely able to pile up dozens of websites and other services, including the most complex, are quite easy to find.'[7] Better territorial connectivity makes it consequently possible to better imagine distributing data storage capacities in local data centres. Thus, SCANI operates three of these directly and occupies two that belong to third parties (Virtua Networks in Auxerre). The cooperative uses them for its internal needs, but plans to create a third-party associative or cooperative structure that could manage these storage spaces for external needs as well. It is interesting to examine the three facilities directly operated by SCANI. The first is a light garden construction, converted into a data centre, which can house three racks (a single one is in service, and a quarter filled). At this stage, the space is not air-conditioned. It is equipped with an undulator. It will be improved with growth as needed. The second is in a former 12 m^2 telecom shed, the property of a mayoralty that lends it to the cooperative. This five-rack space is completely equipped (with electricity and air-conditioning). A half-rack is occupied today. The potential can be seen here of the conversion of a small-scale telecom structure, probably vacant on many territories, notably in rural zones. The final facility is a 30 m^2 technical room in a former military building used as an optical head for a hyper-local FTTH network and a data centre with a maximum capacity of three to four racks. For Bruno Spiquel, when data is hosted locally, for a close community of users, 'the consideration of the energy issue is immediate, but not necessarily "to help reduce the climate impact", often quite simply for questions of electrical and/or financial sizing or, more marginally, available space'. The modesty of these actors' resources therefore leads to a frugal approach to the equipment, layout and functioning of these local data centres.

Another example is Aquilenet (Figure 15.1) in Bordeaux. Aquilenet is an associative ISP that wishes to provide a neutral and open Internet. It proposes hosting[8] as well, and was also able to install its own associative data

---

6 'Regardless of the use made of the connection or the person who subscribes to it (individual, company, etc.), the rate is identical: €30 tax included per month', SCANI website, Available from: www.scani.fr [Accessed 26 February 2025].

7 Email exchanges with Bruno Spiquel, activist at SCANI, Paris, November 2018.

8 Hébergement au format libre (alias Datacenter associatif), Aquilenet, Available from: https://atelier.aquilenet.fr/projects/services/wiki/Hergementlibre [Accessed 26 February 2025].

Figure 15.1: Storage architecture at Aquilenet

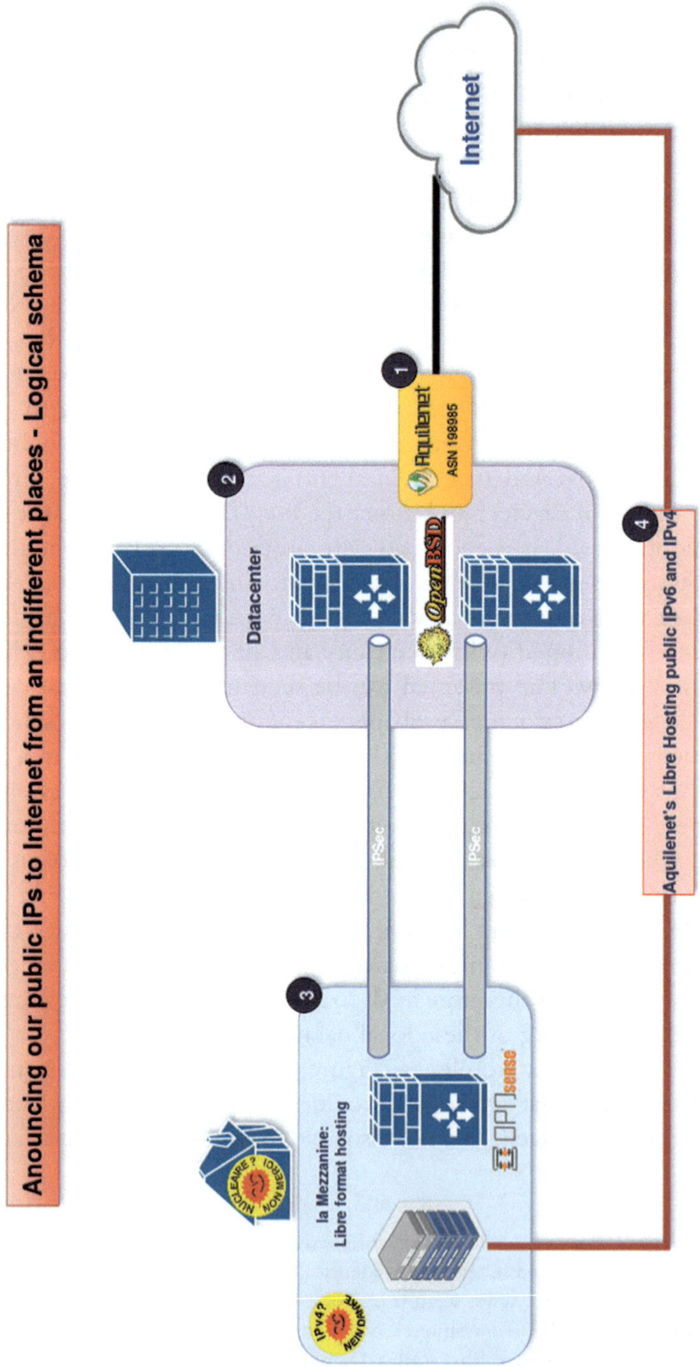

Source: Aquilenet Internet site

centre with the free format ('the Mezzanine')[9] in the autumn of 2017 on premises where the electricity is provided by a supplier that only proposes renewable energies.

The monthly price for hosting on the Mezzanine is between €5 and €10 for a small machine (Next Unit of Computing [NUC] type, netbook, laptop, Network-Attached Storage [NAS], Raspberry pi, etc) and between €15 and €20 for a PC (consumption of 100 W maximum for an average consumption of 30 W).[10] Aquilenet also hosts its servers in a professional data centre (COGENT). The rate is €100 a month per rack unit.

In Toulouse, Tetaneutral is an ISP, Web host and operator in the form of a not-for-profit association, which has been hosted since 2011 at Mix'art Myrys, an art and culture structure, in a former hangar in a Toulouse industrial zone of 6,630 m^2 in area, made available thanks to the public subsidies of local communities (City of Toulouse, Toulouse Métropole, Haute Garonne Department, Midi-Pyrénées Region and the Regional Direction for Cultural Action). The site is divided into 4,200 m^2 of workshops, 430 m^2 of offices and 2,000 m^2 of exterior spaces, but Toulouse Métropole decided to sell it in early 2021. The Tetaneutal servers were therefore moved to new premises in the Minimes district, a space that previously housed a printing plant and that now offers studios and workshops for artists and craftsmen.

At Mix'art Myrys, the association had two computer rooms, on the upper floor of the building, near the offices, with a connection to a neighbouring data centre operated by COGENT. The association had initially rented a rack unit in a bay, with a transit contract and had the right to drill a hole to bring in a fibre cable from the bay to Myrys (the transit continues, but via another fibre cable). Moreover, Tetaneutal co-operates a room in classic data centres, operated by Fullsave, which proposed an 'associative' room to them that was difficult to air-condition, outside the ultra-secure zone. It was, in reality, a residual space in the building that, by adding a partition, a door, an electricity inlet and a fibre inlet, could provide a space at little expense. The association pays an inexpensive rent and a low electricity bill. Tetaneutral refuses to host in those machine rooms that consume too much electricity, and raises its members' awareness on electricity consumption questions. It is very interesting to note here a complementarity between a classic and an associative data hosting centre, as well as the optimization of residual spaces in the building. Tetaneutral has recommended rates, which implies there are de facto social rates. They are the following: hosting of a virtual machine: €10/month;

9 La Mezzanine: histoire d'un centre d'hébergement associatif au format libre!, Aquilenet, Available from: https://www.aquilenet.fr/actualit%C3%A9s/la-mezzanine-histoire-dun-centre-dh%C3%A9bergement-associatif-au-format-libre/ [Accessed 26 February 2025].

10 Hébergement, Aquilenet, Available from: https://www.aquilenet.fr/services/h%C3%A9bergement-serveur/ [Accessed 1 December 2018].

hosting of a physical machine – small machines (NUC, Brix, Zotac, netbook, laptop, nettop, NAS, sheevaplug, raspberry pi): €5 to €10/month; and PC: €15 to €25/month (€20/month for a consumption of 100 MW maximum, with an average of 30 MW).[11]

These new small-scale infrastructural locations are not a return to the computer closets of companies in the 1990s, but a potential relocation of the Internet as close as possible to users, in the heart of cities, with a frugality and technical sobriety logic, but also of technical adjustments and autonomy. They are thus redefining the idea of public facilities on the territories, and anticipate a new, more local, more distributed geography of the digital, whose energy impact, however, should be objectively compared to the centralized and industrial system. Does the frugality aspect of the computer equipment production phase (this equipment is often recovered, reconditioned) compensate for the possible weakness of the functioning and energy consumption phase? This would need to be measured in order to be verified.

The technical competencies developed in these actors are at a high level, but no wishes have been expressed to scale their infrastructures as Guifi.net has proposed in Spain, or to transform into commercial companies. The objectives tend more towards proliferation and autonomous developments. Certain people could criticize these associations for a lack of professionalism, if professionalism equals entering the classic data centre market. Their objective, however, is not to host all the data of CAC 40 companies,[12] but more a reasoned storage of individual data and small companies favouring digital frugality. These structures represent an alternative for a plural digital system. They can exist next to the market's actors and be deployed thanks to public support linked to digital public spaces, for example, in third places or with any other budget raising the questions of access to the digital. The density of this network's actors could then increase and further connect the territories.

Calafou: technological sovereignty and self-management in Catalonia

Calafou is a former factory, transformed into a community which is self-managed, in the form of a cooperative, in which a specific reflection on the reappropriation of digital tools is being carried out. A hackerspace and autonomous servers are hosted in the premises. Calafou's objective is

---

[11] Adhérer, Tetaneutral, Available from: https://tetaneutral.net/adherer/ [Accessed 26 February 2025].

[12] The 40 most significant stocks among the 100 largest market caps on Euronext Paris.

to 'generate production, technological and housing alternatives that are sustainable and respect the environment'.[13] The former industrial colony was collectively purchased, in a state of advanced deterioration, and extends over 2.8 hectares, on which are deployed 'spaces for industrial, artisanal and creative activity, shared infrastructures for meetings and various activities, and even decent housing. All of them are accessible at fair prices, notably below market prices and designed to permit self-management of the project/ place'.[14] It is in the approach of 'social, technological and political innovation based on accountability and cooperation'[15] that several projects and places linked to the reappropriation of technological tools were developed with the idea of taking back sovereignty, and notably autonomous Anarchaerver server projects.

Spideralex (also called Alexandra Haché), a sociologist with a doctorate in social economics, is also an activist connected to Calafou. She stresses that:

> Feminist servers have been a subject of discussion and a group of technical-political practices developed by networks and groups of cyberfeminists and transfeminists interested in the creation of a more autonomous communicational infrastructure. They want data, content and the memory of feminist groups to be easily accessible, preserved and managed. This leads to gaining autonomy in the access and management of our data and our collective memories.[16]

Spideralex crosses feminism and the question of technological sovereignty. She published, under Creative Commons, two volumes on technological sovereignty, one in 2014 and another in 2018. The question of autonomous servers is notably developed in these volumes. The Anarchaserver project incarnates this positioning. It is a self-managed computer server, located in a 12 m² room at Calafou, founded in 2013 during the TransHackFeministas days. It hosts cyberfeminist projects and the digital data of collectives, archives and sites. Furthermore, there are other similar initiatives in Brazil (Vedetas, C14ndestinas) and Mexico (Kefir), for example. Spideralex created the feminist collective Donestech. She took part in the creation of Anarchaserver and is also interested in transforming it into a living and comprehensible physical space:

---

[13] Calafou, Available from: https://archive.calafou.org/en/content/about.html [Accessed 8 May 2022].

[14] Calafou.

[15] Calafou.

[16] Spideralex, Soutenir ce qui nous soutient: faire de l'infrastructure féministe, Available from: https://alexandria.anarchaserver.org/images/7/71/388-Texte_de_l%27article-891-1-10-20210702.pdf [Accessed 26 February 2025].

> We found a dilapidated place and we renovated it so that it would become a space for living and meetings and that made the server a physical entity. When you arrive in this space, you have to log in, you physically write your login somewhere, there are mirrors, like online, you're going to have mirror servers, there's a firewall except that it's just a portable stove in which wood is burning to provide heat … It's fun, it's a way of creating a habitable and appropriable space for the cyberfeminists who need a place to sit down and concentrate. And then, this creates a common, a resource to be shared, it's a joyous way to show what a virtual machine consists or can consist of.[17]

Today, at Calafou, the Arduinna laboratory is a 40 m^2 collective space for developing projects on electronics, hardware, information technology and philosophy. This hackerspace is a continuation of previous projects like Pechlenda, Biolab and Hacklab. Anarchaserver (Figure 15.2) is still hosted at Calafou, and concretely illustrates the question of the reappropriation of digital technologies connected to political, surveilled and/or repressed battles in certain countries.

Internet, infrastructural commons supported by a renewed governance

In this reappropriation dynamic, there is, whether it concerns energy or the digital, the desire to organize a common, whose citizen support (inhabitants, local economic actors) can be increased by institutional and/or municipal backing. Creating a place for these actors, who are small in scale and with a low number of users, in a French digital landscape that is not very diverse today, proves to be important, to the degree that a more local, better-informed management, handled by its own users, can favour more sober, economical practices adjusted to meet needs. The alternative digital actors are working to redefine the Internet as a common, updated to deal with the situation today, which is now both very inegalitarian faced with Big Tech, but also very fragmented. There are Internet user communities with different practices and interests: that of the alternative Internet is meshed and inventive and calls on solidarity. It is found in a great many countries in the world. The users converge at dedicated festivals or events such as the Pas Sage en Seine festival (Société, Liberté, Internet), which has taken place for over ten years in France. There are local communities as well, which help

[17] For more information, see Pas d'internet féministe sans serveurs feministes, Panthere Premiere, Available from: https://pantherepremiere.org/texte/pas-dinternet-feministe-sans-serveurs-feministes/ [Accessed 26 February 2025].

Figure 15.2: Homepage of the Anarchaserver site

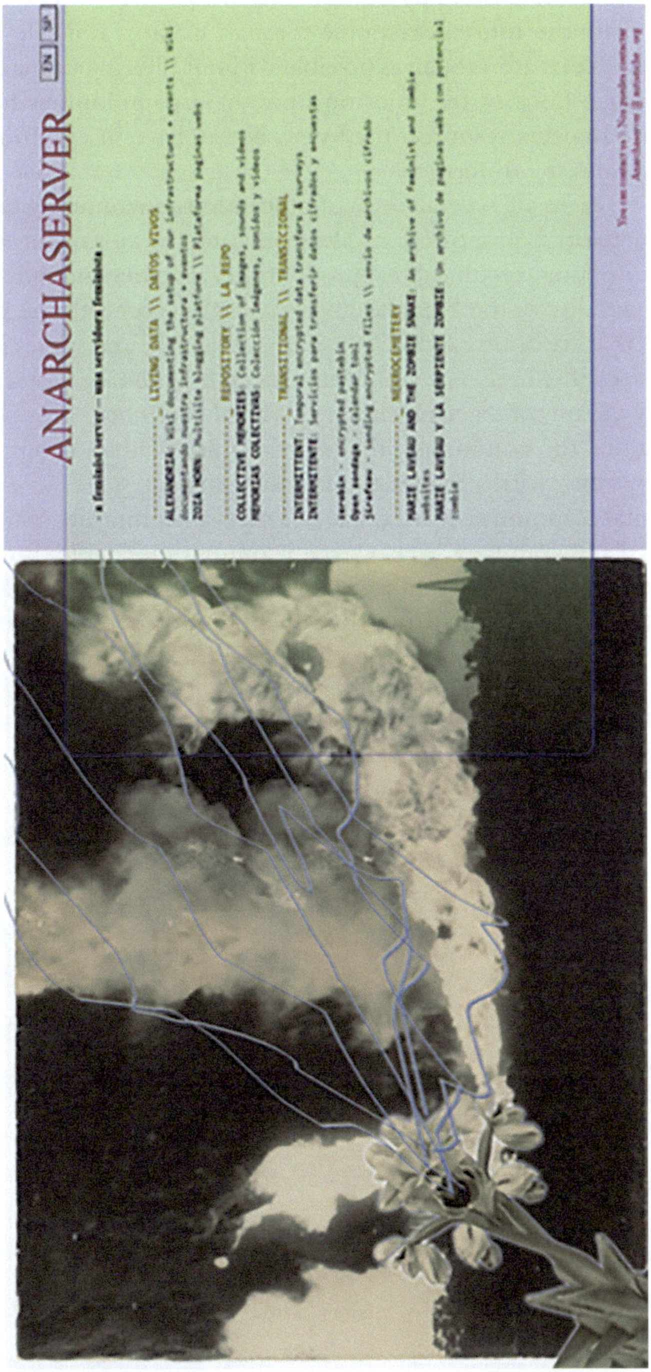

Source: Anarchaserver.org

one another and intervene in remote territories like rural Catalonia with Guifi.net or in the Burgundy countryside with SCANI. Their approach is to best share the Internet resource through making available the digital infrastructures that are as open as possible without filtering or manipulation. As for the questions of the rules on how these communities function to manage the Internet resource, they vary, but all have in common the aim of a neutral and free Internet.

Moreover, there are several kinds of motivations surrounding community Internet networks: proposing an alternative to the dominant actors that provides autonomy, technical reappropriation capacities and independence, but also neutrality of the Net and liberty; offering access to the Internet in territories that are poorly or not connected, like the rural spaces in Europe; and permitting the resilience and continuity of functioning of spaces that are potentially victims of extreme climate events. More rarely brought forward is the idea that the community digital networks could also comprise more cautious options, with a lighter environment impact.

The digital common coexists today in a very small Internet space with the commercial and dominant Web. Certain communities could, however, rely on this renewed digital common to develop public/common partnerships in order to offer a relocated Internet. Despite its modest role, it is a solid option for decentralized approaches, favouring frugality and rethinking time in another way.

16

Resilience and Inclusion: Systems Faced with Disconnection

Certain territories and communities had to find alternative systems to connect in so-called white zones or to not be disconnected to the Internet after extreme climate events like storms or floods, notably causing power cuts. Their approach is therefore a matter of decentralized or distributed techniques, but more to be integrated or maintained than to protest and transform.

Red Hook: from digital resilience to the micro-energy grid

Red Hook is a neighbourhood in Brooklyn, New York, a peninsula and therefore extremely vulnerable to flooding. It has around 11,000 inhabitants, most of them young with modest incomes, and with over half living in social housing. Questions of social and environmental resilience converge on this territory: floods, Internet disconnections and power cuts are frequent, while the local population's access to the digital network remains less developed than in the rest of the city, reducing access to jobs, knowledge and training.

The Red Hook Initiative (RHI), created in 2002, would attempt to tackle the problems, gradually starting in the 2010s. In 2010, this not-for-profit structure moved into a dedicated building, notably to reach more residents in the neighbourhood. RHI realized that to maximize its impact, it also had to build a local digital community. The question of digital inclusion thus became strategic starting in 2011. With the help of the Open Technology Institute and a student from the Parsons School of Design, RHI began to provide free access to the Internet through Wi-Fi, managed by the community's inhabitants (Figure 16.1), and more specifically by young people in reintegration who are trained to manage and repair the network:[1]

[1] New America Foundation, 'Case Study: Red Hook Initiative Wi-Fi & Tidepools', New America, Open Technology Institute, 1 February 2013, Available from

Figure 16.1: Repair of the meshed Wi-Fi network node in Red Hook after Hurricane Sandy

Source: © Alyx Baldwin, 2012

the digital stewards. The network hub was successfully installed in the RHI building to connect the neighborhood, via a Ubiquiti nano-station on the roof and a router in the building, connected by ethernet to the centre's modem. The access to Wi-Fi also provided access to the RHI website, to a forum, in order to enable the involvement of the inhabitants in local projects. In March 2012, another nano-station was installed on the roof of a residential building in order to increase the network's coverage. This nano-station was at first not connected to the Internet, but to a GuruPlug server, a small, low-consumption server functioning on Linux. It then hosted a local portal of the same type as the RHI website. The network is therefore ultra-local. It functions using Commotion software that makes possible the creation of mesh networks. Moreover, specific digital tools for and by the community were developed, through workshops, notably the following two applications: Where's the B61 Bus (an application in real time on the locations and arrival times of the B61 bus; see Figure 16.2), and Stop & Frisk survey (an application so that the

https://www.newamerica.org/oti/blog/case-study-red-hook-initiative-wifi-tidepools/ [Accessed 26 February 2025].

Figure 16.2: Sign advertising the 'Where's the B61 Bus?' Web application available through the local Red Hook Wi-Fi network

Source: © Alyx Baldwin, 2012

inhabitants can obtain information on their interactions with the police and improve local security).

But it was when Hurricane Sandy caused flooding throughout the entire neighbourhood in 2012 that RHI revealed the strategic potential of a local Internet network. Thanks to the local network, but also to the backup infrastructures of the RHI building (generators), the building became an open refuge 14 hours a day for 24 consecutive days. The site was able to offer electricity, water, heating and the Internet, thus facilitating contact for emergency services, as well as between people. In the first days after the storm, nearly 300 people were connected each day. The local Internet network continued to function. Through the question of inclusion and access to the Internet, it was also that of resilience faced with floods and power cuts that was consequently addressed.

Since then, the network has been largely extended. In 2015, an initiative led by other neighbourhood actors[2] to set up an energy micro-grid was awarded a prize from the New York State Energy Research and Development Authority (NYSERDA) to conduct a feasibility study, with the objective of improving the neighbourhood's autonomy faced with the power cuts and fragility of the grid in Brooklyn. It concerned thinking together about the development of a micro-Internet network and a micro-electricity grid and to explore how they could interact. The project was not selected for phase 2 of NYSERDA financing, so it was never concretized. It planned to supply electricity, heating and hot water through a gas cogeneration plant and solar panels with, as backup, diesel and gas-powered generators, to nearly 2,800 apartments (6,300 people) owned by the social housing landlord, the New York City Housing Authority, but also emergency shelters in the event of a natural disaster: the RHI building, an IKEA and other critical sites. The combination of the micro-electricity grid and the citizen Wi-Fi, as can be seen on the map in Figure 16.3, would have been an innovative coupling for the neighbourhood's autonomy and resilience, given the projections of recurring intense climate events brought forward by the Intergovernmental Panel on Climate Change reports. This is all the more relevant as the micro-grid itself needs an efficient and robust data communication network to function, in the same way as the communication network needs electricity. In the event of a power cut on the micro-grid itself, this project also foresaw a backup system for each mesh gateway node with a photovoltaic panel and a battery system. Since 2014, Red Hook Wi-Fi has moreover been developing solar energy for its mesh network, through subsidies from the New York Economic Development Corporation, for Rise, a New York City program. The agency has also helped develop five other resilient local Wi-Fi networks in East Harlem, Hunts Point, Gowanus, Far Rockaway and Sheepshead Bay, with the support of local institutions, the participation of residents and the contribution of the Resilient Communities program of the New America Foundation.[3] Social integration, energy and digital resilience, access to the Internet: the Red Hook project brings together several major dimensions and proposes an interesting model for all sorts of communities.

---

2 Smarter Grid Solutions (technical engineering), IMG Rebel (consultant on financing), RH-NYR-CRP (stakeholder lead and community co-lead) and FBCB6 (community co-lead and project admin).

3 For more information, see Resilient Communities, New America, Available from: https://www.newamerica.org/resilient-communities/ [Accessed 27 February 2025].

Figure 16.3: Infrastructural mediation map

Source: François Huguet

W4C: a global digital inclusion program

In a more globalized digital inclusion approach, the Internet Society and the Digital Empowerment Foundation program called Wireless for Communities (W4C) has been deployed in India since 2010. W4C uses wireless technologies, Wi-Fi equipment at low cost and unlicensed frequencies – 2.4 GHz and 5.8 GHz – to create wireless networks set up and managed by local communities to democratize access to the Internet and make it accessible to everyone. W4C started as a community network pilot project in 2010 in a rural Indian village, Chanderi, a community traditionally associated with textiles in the state of Madhya Pradesh. This program, notably supported by Microsoft, Tata Trust and the Vodaphone Foundation, then followed several development stages: with trainers called the 'barefoot engineers'; in other Indian states; with a program targeting women; by connecting local networks to one another; and by bringing the Internet to local governments, school and public services. The barefoot engineers are people who train the local communities to set up, manage and repair citizen Wi-Fi networks, but who also train other trainers in villages or regions. The program has 146 access points today in 38 districts in 18 Indian states. One of the objectives is the integration into globalized commercial channels and local economic development.

The data mule and the sneakernet: the low-tech Internet

Delay or Disruption Tolerant Networks – DTN[4] – are developed in countries where the electricity service is not deployed or reliable enough to permit a continuous Internet. They are consequently subject to disconnections and very long latencies. Unlike the always active traditional networks, these DTNs are based on asynchronous communication and intermittent connectivity.[5] They overcome the problems of intermittent connectivity and long delays by using a message switching system. The information is transferred from a storage location on a node to a storage location on another node, along a route that finally reaches its destination. These networks are notably used for space communications. Based on the principle of mules and autonomous messages (bundles), they are established using equipment with large storage capacities in order to ensure the routing of data despite

[4] Unlike traditional Internet routers, which only store incoming packages for a few milliseconds on memory chips, network nodes with a delay tolerance have persistent storage (such as hard drives) that can contain information for long periods.

[5] These networks have their own specialized protocols, superimposed on lower protocols, and do not use the Transmission Control Protocol.

network disconnections. Different deployments have been proposed for animal surveillance (zebranet network) or the provision of Internet access to populations far from network infrastructures through mobile equipment (snowmobiles, bus networks or mail carriers on motorcycles). Consequently, transportation infrastructure can replace a wireless Internet link. The Daknet, Saami, DieselNet and KioskNet networks, for example, use transportation equipment (buses, motorcycles) as data mules in many villages and cities without network connectivity. Each 'mule' vehicle is equipped with a computer, a storage peripheral and a mobile Wi-Fi node, and a stationary Wi-Fi mode is installed in each village or town. The emails sent or the requests for Web pages are stored on local computers in the village until the arrival of the bus or motorcycle that will transmit outgoing and incoming data.

These networks with a delay tolerance (used in the mid-1990s with the Unix-to-Unix Copy protocol) can function with an intermittent energy supply and combine well with renewable energy sources: solar panels or wind turbines can only supply network nodes when the sun shines or the wind blows, consequently eliminating energy storage needs. A large number of Internet applications can be adapted to intermittent networks, such as email, electronic form-filling, blog software, browsing on the Web, interaction with e-commerce sites, downloading files (even voluminous ones) and social media.

Real-time applications such as Internet telephony, continuous multimedia broadcasting, live discussions and videoconferences are impossible to adapt to intermittent networks, which solely provide an asynchronous communication, but it is perfectly possible to send and receive delayed voice or video messages. And even if continuous multimedia broadcasting is not possible, downloading music albums and videos remains achievable. Furthermore, these files can be 'transmitted' by the lowest Internet technology possible: a sneakernet. In a sneakernet network, digital data are transmitted 'wirelessly' by using a storage support such as a hard drive, a thumb drive, a flash card, a CD or a DVD. Before the Internet's arrival, all computer files were exchanged via a sneakernet, using a tape or disk as storage support.

Like François Huguet's infrastructural mediation map, the actors (FFDN, Guifi.net, alternative Internet Service Providers, for example), technologies (Commotion, Bluetooth, etc) and places described here are part of a much larger geography that is densifying and proliferating, unlike the major digital actors that are merging and consolidating into centralized and hierarchized units. They are very closely linked to a political vision of the world, maintaining democratic initiatives (Indignés, Occupy, Nuit Debout) and the technical appropriation of the tools that model our lives.

17

The Ecological Footprint of Alternative Digital Infrastructures

For distributed and decentralized solutions, the absence of backup infrastructures (backup generators, batteries) and large air-conditioning systems, but also electrical equipment like undulators, reduces the expense of the initial material (production), integrated energy used and energy consumed in the event of a breakdown. On the other hand, a breakdown on a distributed storage system can generate the mobilization of more machines on the peer-to-peer network and cause additional energy consumption.

A question that is central to these alternatives is the energy impact of distributed storage and data centres of proximity. Several hypotheses are possible, but there is very little research on the subject at this stage to compare decentralized and distributed systems to today's dominant, centralized and globalized system.[1] Nonetheless, several indications seem to be in favour of alternatives.

For distributed storage as for local data centres, visualizing the physical spaces occupied by the servers favours a better-adjusted functioning: the storage capacities are not oversized, the servers are not on standby like zombie servers ... These storage modes avoid the drawbacks of dematerialization – that impression of having access to infinite storage spaces, without any impact on the environment. Storage is also carried out more locally (not on continents far from the users) and the data travel less on communication networks; we can therefore suppose that less energy is used on the network. These initiatives of digital technology reappropriation

1 A research track of the Groupement de Services EcoInfo explores questions on the energy impact of storing one's data on a local hard drive rather than on the Cloud.

are in reality following a movement and dynamics comparable to those that can be observed around renewable energies by cooperative and citizen groups (sometimes supported by communities) in Europe. Laure Dobigny, a sociologist-anthropologist specializing in techniques and technologies, studied the effects of relocating renewable energies on users' behaviour, through the analysis of autonomous rural communes. She puts forward that energy management that includes and involves, at one moment or the other, the inhabitants in the process, participates in a dynamic of restraint.[2] This is what Dobigny observed on rural land, where the installation of smaller-scale technical systems made possible, by modifying uses, a reduction in consumption.

Moreover, the 'IT Refresh' processes that are performed every two years in colocation and Cloud data centres are probably less frequent in the case of individual machines or associative storage. Here, too, we could put forward the hypothesis of an economy of materials and embodied energy at the computer equipment production stage, and therefore less production of computer waste as well. This limitation of the consumption of new computer equipment corresponds to both the recommendation of the FING IDDRI WWF white paper[3] on 'extending the duration of computer equipment guarantees to five years' and the recommendations of the report by The Shift Project on energy and the digital published in 2018.[4] It is also found in law no 2021–1485 of 15 November 2021, which aims to reduce the environmental footprint of the digital in France. From an economic and social viewpoint, data centres are proximity services, hosting is accessible for local activities and the small- and medium-sized enterprise fabric is rich, notably in France. A training offering for the maintenance of digital infrastructures is often proposed and can be a springboard for job creation.

As for the measurement of spatial and territorial impacts, this local and decentralized model has not been accompanied by urban and spatial concentration phenomena as observed with a centralized model. Consequently, the following are avoided: urban heat island effects, and pressure on the local electricity system and on land resources. A local and decentralized system also favours greater resilience faced with extreme climate events, compared to that of very centralized large-scale infrastructures. The

---

2 Laure Dobigny, *Le Choix des énergies renouvelables. Socio-anthropologie de l'autonomie énergétique locale en Allemagne, Autriche et France*, thesis under the direction of Alain Gras, Université Paris 1, 2017.

3 *Libre blanc Numérique et Environnement. Faire de la transition numérique un accélérateur de la transition écologique*, FING IDDRI WWF.2018.

4 Hugues Ferreboeuf (ed), *Lean ICT, Pour une sobriété numérique*, The Shift Project report, October 2018.

weakness of alternative storage could be summed up as its potentially lesser availability, which implies that users will be very tolerant about breakdowns; but also a non-automatic redundancy of the data storage. It is therefore incumbent on the user to anticipate these. Criticism as to less data security in these alternative systems seems challenged by all the scandals on data leaks surrounding the large global digital operators like Facebook and Yahoo. The recurrence and strong growth of IT attacks represent in reality a risk for all digital infrastructures.

Despite the high-tech nature of industrial data systems that aims to protect them from heatwaves, possible intrusions or IT attacks, they are in general systems that provide fortification, but have little regard for territory. If, for example, the Interxion data centre on the rue Rateau in La Courneuve was renovated so that it could continue to cool the servers despite heatwaves, hot air would continue to be discharged into a densely built territory, in which the populations experience more health problems than in the rest of the Île-de-France region. A power cut on the electricity grid for an entire data centre district would furthermore involve, as in New York in 2012 with Hurricane Sandy, lines of oil tankers supplying fuel to the backup generators to ensure service continuation, thus locally generating air pollution, road safety problems, congestion, noise, etc. Unlike these large data centres, a decentralized spatial system reduces the risks of a massive service interruption and the necessity of implementing backup actions with negative local impacts.

Thus, storage distributed in computer terminals and small data centres of proximity comprise a possible solution for reducing the spatial (notably land take) and energy effects of storage, while being a training and local economic development lever. Additional studies are needed on this subject comparing the services rendered. Consequently, the electricity provision of the infrastructures and the possibility of (self)consumption (or not) of renewable energies must be taken into account.

If, today, hosting, storage and data processing remain mostly in the hands of commercial operators, civil society operators are developing, and public actors are also rolling out their own infrastructures. The development of public data centres and the social and solidarity economy could consequently constitute a perspective permitting a more coordinated relationship with the territories, accompanied by requirements for pooling and energy savings.

These examples illustrate the diversity of resources to imagine several degrees of connection, connectivity and resilience faced with a given geographic isolation or one caused by floods, as in Red Hook. If the world of banking, finances, advertising and streaming, which represents a large part of Internet traffic and the computing and storage capacities of data centres, are probably not ready to turn to public or more distributed structures, there are, however, opportunities here for civil society, associations and the social and solidarity economy in order to regain control of this common good – the Internet. The scale and modalities of these infrastructures' governance are major societal subjects at a time when the social ideal of the large infrastructure as a public service edifice, which combined economy of scale, technical stability and quality service for the largest number of people, has been destabilized since the 1990s by the private capital of the liberal economy.[5] The centralized technical object is prey to new assemblages and displacements of what comprised its value. The overhauling of public service based on commons,[6] remunicipalization and deprivatization movements on municipal or regional scales are multiplying in Europe, as well as digital services, as in Stockholm and New York. The support of the territories in access to a free and open Internet is as indispensable as a broader reflection on the *data centre* object and the associated digital system; to better measure the environmental impact of the technical choices with regard to the expected social added value; and to shift towards more reasoned, sober and degrowth digital practices; in short, towards more technical democracy. The motivations driving these alternatives are, on the one hand, integrationist, with the objective of connecting isolated territories to worldwide commercial channels and globalized consumption modes and, on the other, secessionist, in order to either offer digital infrastructures for a free, neutral, uncensored Internet, without being dependent on large private telecom and digital groups, without any prioritization of traffic, censoring or filtering of contents, or to reach an asserted national sovereignty. Some of these systems rely on a technical appropriation by the communities concerned, aiming at opening the black box of digital infrastructures at least for the networks. The connection between the reduction of the environmental impact of the digital and the frugality that these initiatives incarnate is not always explicit in the discourses concerning these projects. Nonetheless, this is indeed the path they are taking.

---

5 Steve Graham and Simon Marvin, *Splintering Urbanism: Networked Infrastructures, Technological* Mobilities and the Urban Condition, London, Routledge, 2001.

6 Benjamin Coriat, Nicole Alix, Jean-Louis Bancel and Frédéric Sultan (eds), *Vers une république des biens communs*, Paris, Les liens qui libèrent, 2018.

PART V

Alternative Futures: Three Possible Digital Worlds

If the technologies described above as alternatives sketch a frugal world in which the digital permits more freedom, communication, more creativity than surveillance, centralization and individualized tracking, the possible futures are being woven in a more complex fashion than that binary face-off, encountering growing climate, environmental, migratory and democratic crises.

What will the Internet of tomorrow be? Reflections, debates and international regulations on the digital lead to a rather centralizing perspective whose trajectory the first scenario proposes to extrapolate. The second scenario asks us to follow another trend: that of the structuring of a decentralized digital public service coexisting with a small-scale infrastructural diversity. Lastly, the third scenario marks a radical break, the end of data centres and the digital world as we know it makes way for distributed networks primarily based on intermittence.

18

New Digital Worlds

Scenario 1: Digital growth and ultra-centralization

The monopolistic strengthening of the digital industrial complex was reinforced over the years. The large dominant companies of the sector encountered no real obstacles, no anti-trust barrier, given how much the native states of these Big Tech companies made them powerful weapons and want to keep them as such: a soft power vector; levers of creating dependence of very many countries via software, tools and data; a Trojan horse in foreign administrations, on worldwide conflicts and geopolitical balances, elections or industrial espionage. The European Union's efforts have not borne their fruit: its influence is irremediably decreasing on the global scene. Big Tech (GAFAM and BATX) is therefore sharing a worldwide market that remains very profitable. These companies are pursuing their growth, Internet traffic has exploded: 200 billion objects will be connected in 2030. The public authorities have observed, powerless, a very large increase in the storage needs and correlatively electricity needs of the digital sector, anticipated to consume 51 per cent of electricity worldwide in 2030.

An increasingly centralized, secure and controlled Internet architecture

This is the culmination of the shift from the horizontal to the vertical model, which had already started in the 2000s. The Internet architecture became a group of networks dominated by ends linked to the financializing of exchanges, the platformization of services, targeted information and the manipulation of minds, surveillance and cybersecurity. The increase in filtering systems and differentiated connection speeds according to the contents marks the definitive end of the Net's neutrality. Its margins are criminalized, the free Internet is tracked. The proprietary architecture model prevails, the networks support machines of an increasingly reduced number of manufacturers for better control of flows.[1] This is the strengthening of the vision of the Internet as an information highway in which a technical and economic model favouring the relations between customers and monopolistic and powerful servers dominates. Internet governance is left in the hands of private economic actors with heavy lobbying by Big Tech that maintains its financial pressure on politics with the reinforcement of the existing monopolies, especially on data. This growth logic favours the centralization and consolidation of the market around GAFAM and BATX, whose capitalistic power dominates and creates the supply. There are very few mid-size actors; the colocation has also been consolidated around a handful of international actors.

LTS spatiality: in the large group of machines

The digital technical system follows the logic of the large technical systems (LTS[2]). There is historically a correlation among the large scale, the economies of scale that it permits and centralization. The continual growth of the Cloud, largely in the hands of GAFAM, took part in the multiplication of very large-scale data centres (50,000–100,000 m^2), while developing a reticular system combining the smallest-scale objects (colocation or not), but also edge computing. Digital zones like Ashburn have multiplied everywhere in the world, close to the large metropolises. Whole economic activity zones have mutated to data storage combining new buildings and transformed buildings. Logistics warehouses and data warehouses are increasingly found side by side, Amazon systematically developing both of them. The domination of e-commerce has had a devitalizing effect in almost all the shopping districts in cities, exclusively keeping dark stores. Shopping centres on the outskirts of cities are themselves being transformed little by little into sites for digital data and merchandise storage, without any visibility to the public except for curbside pickups.

All the Big Tech companies have become major actors in urbanism and regional development with the multiplication of ultra-connected urban districts. If Google's urbanism subsidiary, Sidewalk Labs, failed at first with the creation of the Quayside district in Toronto in 2020, the GAFAM companies have heavily invested in the real estate and urban management

1 Francesca Musiani and Valérie Schafer, 'Le modèle internet en question (années 1970–2010)', *Flux*, 3(85–6), 2011, pp 62–71.

2 Joerges Bernward, 'Large technical systems: concepts and issues', in T. P. Hugues and R. Mayntz (eds), *The Development of Large Technical Systems*, Frankfurt, Campus, 1998, pp 9–32.

markets, notably with public spaces and sidewalks, playing on the decreasing tax resources of public administrations.

The world is exploited by the surge of data as far as the most rural boundaries, in which agricultural land takes have become complexified in the spatial developments mixing gigantism and factory-farms, data centres and energy production. Immense production hubs have been deployed on the model of Agriport in the Netherlands, with, nearby, the development of renewable energy or biomass plants like ezCloud (a Cloud computing project that uses methane from farms). New agricultural-digital territories have emerged with the systemization of heat recovery from greenhouses, notably in the Centre Val de Loire, Brittan and Nord-Pas de Calais regions, in which normal temperatures have remained below 45°C.

In the urban zone, the arrival of spaces of networks at two speeds has restructured the city on the principle of 'splintering urbanism' identified by Graham and Marvin.[3] A typical case of the 'secessionary network space' or 'first-class network-spaces', this phenomenon reinforces the social-spatial fragmentation of premium, second-class or third-class spaces. Whereas in the most precarious districts, a return of the thumb drive to exchange data has been noted, the most affluent neighbourhoods and the central business districts make use of infinitely small latencies.

Rare sanctuary white zones are reserved for the wealthiest protected from the electromagnetism of 8G. Outside, on several fronts, imperialistic wars for rare metals continue. In the metropolises and dense urban centres, vertical data centres or micro-data centres are inserted into many empty plots, as well in classic real estate projects (housing, commerce, tertiary, cultural) in order to meet the needs of the Internet of Things (IoT) (low latencies, permanent relays). In Île-de-France, the Société du Grand Paris has notably favoured the creation of 26 data centres in the new metro stations (buried, semi-buried). Data centres are also being housed in many vacant tertiary buildings, emptied by repeated pandemics and the widespread practice of remote work. The development of a nuclear plant-data centre on Mars, in orbit and on the Musk space station are underway.

Energy: the headlong rush

Google and Microsoft have become involved in energy production and distribution, evolving into the sector's major actors with the development of large-scale solar plants and wind farms. They are also specialized in

3 Stephen Graham and Simon Marvin, *Splintering Urbanism:* Networked Infrastructures, Technological Mobilities and the Urban Condition, London, Routledge, 2001.

the sale of micro-grid solutions and have become the major actors in the development of individual and collective self-consumption (notably in the United States, but also in Europe).

This scenario marks the victory of the green growth ideology and the large technical systems vision. The large scale has favoured a specific development mechanism propitious to its own growth; a consumption of large quantities of fossil or nuclear energy; maintenance of unequal ecological trade (extraction of rare earths and manufacturing in countries where labour laws and environmental regulations are absent); the production of heavy CO_2 emissions or 'final' waste; the complexity of groups and subgroups functioning in a network, often opaque and managed in a very centralized fashion by experts; and control and regulation of flows in real time. The electrical-digital system ensures the technical and cultural permanence of infrastructural entities that are the foundation of the expansion of large worldwide economic powers. There are economies of scale and financial but not energy savings: no decrease in global consumption is observed for the sector despite the heavy use of renewable energies. Among the rebound effects, a global increase in the sector's energy consumption and CO_2 emissions is observed. This scenario marks the failure of the control of CO_2 emissions and a forecast planetary rise in temperature of around + 2.5°C by 2050.[4]

Scenario 2: Stabilization of the digital technical system and infrastructural diversity: quest for a difficult resilience

In this scenario, the digital technical system aims to stabilize and control its consumptions with the development of an infrastructural diversity favouring a two-speed digital development, on the one hand, Big Tech and, on the other, smaller-scale infrastructures that are more collaborative, notably supported by public investments.

Alternet: a more hybrid and distributed network architecture

The digital technical system is diversified in terms of network architectures, infrastructures and contents. It accepts more heterogeneity, coexistence

[4] The annual increase in greenhouse gases (GHG) has been continuous. In 2025, the world emitted almost 6 per cent more GHG than in 2015. See: Simulation de trajectoires d'émission compatibles avec le budget carbone +2°C, The Shift Project, Available from: https://theshiftproject.org/wp-content/uploads/2017/12/note_danalyse_les_indc_et_le_budget_carbone_the_shift_project_0.pdf [Accessed 27 February 2025].

and hybridizations. Just like China and Russia[5] have developed their own Web services and applications (designed for domestic control and national sovereignty) alongside the World Wide Web, Alternet proposes services that are alternatives to Big Tech, notably through Framasoft in France, as opposed to the aim of censorship of the Russian and Chinese systems, but also from the viewpoint of local and citizen sovereignty on a free and open Internet. Following the debates on the exponential consumption of bandwidth and the energy effects of the digital, regulations at the European and national scales were voted in favour of the taxation of 'Net Guzzlers'. The result was to reduce the weight of videos and photos exchanged, Internet sites and advertising, on the one hand; and, on the other, to favour local data transmission through the growing use of peer-to-peer systems, between users, but also more data storage on user terminals (computers) instead of the constant use of data on the Cloud (for example, Spotify, Deezer), which reduced in passing the costs of 'peering' the Internet Service Providers.

In many European countries, national public actors emerged for data storage and processing to limit the monopolistic abuses of a handful of private actors. After several failures in having this public actor of data centres (decentralized in the regions) emerge, the state completely changed its strategy: it made its invitations to tender and its allocations to local communities dependent on the use of this public interest Cloud for all the activities of companies and communities. Paired with extremely advantageous rates and a massive communication campaign, this strategy was finally crowned with success.

More diverse and better integrated data centres

In this scenario, there is a drop in the development of data centres, notably those on very large scales (the hyperscales) in rural zones because their electricity consumption is too great, their land impact too heavy and energy pooling difficult. There is also the question of climate and terrorist risks that favoured the search for resilience and a smaller-scale redundancy. Small data centres have multiplied everywhere because the quality of the network permits it, and their spatial integration is easier on this scale. These storage and interconnection spaces sometimes become third places for digital mediation, for learning as well as for Internet resources and contents, as well as its technical functioning and its infrastructures. The users' tolerance for breakdowns and disconnections is greater because they understand the

[5] Limonier Kevin, 'Internet russe, l'exception qui vient de loin', *Le monde diplomatique*, 2017, Available from: https://www.monde-diplomatique.fr/2017/08/LIMONIER/57798 [Accessed 27 February 2025].

reasons for them and the implications. The de-digitalization of a large number of strategic activities has reduced the number of data centres. Information campaigns linked to public health issues on the dangers of ultra-connection and the negative rebound effects of the digital on the economy (destruction of jobs, desocialization, externalization of the benefits) have borne their fruit.

Governance is more local for essential services, and communes, public inter-municipal cooperation structures and regions have strongly taken hold of this data storage and processing industry and of those possible contributions for local energy systems (with shared biomass plants, heat recovery, etc). This ecosystemic concern has been favoured by new types of hybrid equipment, and data storage and computing are associated with citizens' Wi-Fi dynamics and micro-electricity grids (as in Red Hook), notably in rural territories where the connection was fragile. The data and energy centres have become new strategic hubs for the territories. In urban zones, the decrease in logistics activities (owing to the more national, local and circular economy, but also due to the drastic fall in international trade following the financial and food crisis in 2024) has favoured the return of local stores and businesses to the city and close suburbs. Many peri-urban activity and commercial zones have become obsolete and have been transformed into digital activity zones, in coexistence with small-scale artisanal and production activities, notably for reuse, which has returned to favour given the inflation levels that have continued since 2022.

Digital development has followed two logics. The first is very structured around communities that have taken control over – since the decentralization laws of 2024 – on this common good – the digital (like the energy and food communities and cooperatives). In an approach that combines data sovereignty and a concern for the ecosystem, a transfer of wealth from the data centre sector to the local economy (production, service, maintenance) was produced, as well as energy savings and a search for restraint. The second logic follows the peer-to-peer model and a topology of the pre-commercial Internet, in which, once again, Internet user communities that develop alternatives to the search giants and online sharing (decentralized social networks, search engines and peer-ro-peer [P2P] contents) flourish. But the model is very fragile because if the P2P models exist and some of the storage has been relocated, the major IT market actors have kept control, despite the crisis, and continue to set their conditions (notably on the Cloud). The large companies continue to solicit, in most cases, the powerful industrialists that dominate the market on an international scale on computing, information storage and traffic management. The balance of the coexistence of public actors and the common and private actors, as well as the local/global ratio, remains very fragile and favours the development of two Internets.

Energy: electricity degrowth, a fragile control, but ambitious public policies

The digital sector was forced to decrease CO_2 emissions by 5 per cent a year[6] between 2025 and 2050, that is, to go back to the global energy consumption (to produce the equipment and make it work) of the sector in 2013 when it was 2,000 TWh (for 3,200 TWh in 2018 and 7,000 TWh anticipated in 2025[7]).

Regulations have made it possible to restrict traffic that is a heavy consumer of bandwidth by limiting certain uses and taxing those that propose them. A whole segment of services (email, blogs, browsing, social media) has become much more economical and programmed obsolescence is now heavily penalized.

The locations of data centres are now defined in compliance with the Regional Climate Air Energy Plans and their territorial variations, which have finally included the digital dimensions. New local 'land, urbanism and energy' and 'degrowth of flows' plans oversee their installation according to the transformation potentials of the territories, heat recovery and connectivity. The interest is to limit their impact and development.

Scenario 3: Digital ultra-decentralization: the end of data centres

Following various collapses, the Internet has become local and intermittent. In a sort of *IT survivalism*, the founding principles of the Internet (horizontal and distributed) have been radicalized. The networks mostly function on the goodwill of communities of interest, and the Internet is only active when energy is available. In the context of globalized pollution and shortages, energy is intended for the maintenance of essential services. A large part of digital uses (Video on Demand [VOD], IoT) has disappeared.

An intermittent and discontinuous architecture

The worldwide frequency of extreme climate events, between heatwaves, droughts and hurricanes, has weakened the durability of the electrical-digital system. Ultra-centralization flaws have not withstood natural disasters. Technical systems have gradually relocated as in the musical chairs of energy priorities, the commercial digital has sunk to the bottom of the list, far below the needs of the worlds of healthcare, mobilities and food production. The

6 Hugues Ferreboeuf (ed), *Lean ICT, Pour une sobriété numérique*, The Shift Project report, October 2018.

7 *Lean ICT* report.

nuclear accident in 2028 caused a digital blackout and led to the bankruptcy of a large part of the private actors.

The long-distance Wi-Fi networks have been deployed with ranges going as far as 100 km.[8] These networks enable users to communicate with one another and to consult websites configured on local servers.[9] Internet users have learned patience and no longer expect instant contents: disconnections, waits and truncated files are common. Thumb drives have made their return and travel physically, including by bicycle, to transport digital data. An application online and offline hybrid system remains an extremely sophisticated and powerful communication network. Even in a disaster scenario in which the much larger Internet infrastructure disintegrates, isolated low-tech networks remain essential local and regional communication technologies. Furthermore, they can obtain content from other remote networks through the exchange of portable storage supports. It seems that the Internet can be as slightly sophisticated or as highly sophisticated as we can permit it to be.

Spatiality: the micro-scale, data mule and long distance

As during the 2020 COVID-19 period, this great slowdown and curbing of the circulation of flows produced a drastic drop in CO_2 emissions. This period of digital chaos and the crash of Big Tech caused deep structural reconfigurations in France, but more largely in Europe, in a widespread raising of awareness. The large data centres became obsolete warehouses, equipments were moved and reused for small-scale and better-integrated storage. There are more low-tech, small-scale servers and machines located at users' homes and everywhere where relay or connection nodes are possible. It is a dramatic change in scale, with the proliferation of light and mobile distributed micro-structures to ensure connection relays with long-distance Wi-Fi systems or data mules. These are vehicles combining energy autonomy, digital equipment and car-sharing. The trajectory of mule vehicles reinforces local commercial linearities and intensifies the occupation of public spaces when they are shared, as well as the development of places for Internet reception, repairs, sociability and work.

Non-connection time would favour IRL (in real life) exchanges and the development of agora (public plaza) locations, discussion, events and

[8] Through modifications of the mandatory access control (MAC) layer in the network protocol and by the use of amplifiers and directional antennas. The longest unamplified Wi-Fi link is a point-to-point wireless connection of 384 km between Pico El Águila and Platillón in Venezuela, established a few years ago.

[9] For an Internet connection, a wide area Wi-Fi network must be connected to a 'connecting node' or a 'gateway node' of the Internet (DSL, fibre or satellite).

conviviality places. The use of the network is limited to certain utilizations considered a priority, notably in healthcare, food and meteorological logistics. The relocation of essential activities and the decrease in work time with 15-hour weeks has turned the use ratios of the digital network upside down.

Decrease in consummation and use of renewable energies in the majority: follow the wind/follow the sun

This scenario is based on an entirely peer-to-peer – mobile – model, itself based on the IoT with a hosting function. Terminals host services for their users and offer these services to neighbouring users in 'degraded' mode. Local digital services become the norm, international services become the exception. The states and especially the Internet user communities have taken control of their services. Certain aspects were explored in Canada with the GreenStar project in 2010 – using the 'follow the wind/follow the sun' principle.[10] The GreenStar project consequently deployed a network that would only function on renewable energies in accepting intermittences on the network.[11] As a preamble, the Canadian project cited a report written by the UK government titled: *How Climate Change Could Ruin the Internet*.[12] In this text, the Secretary of State for Environment, Food and Rural Affairs, Caroline Spelman, stressed that the rise in temperatures and climate change was going to have a strong effect on the reliability of digital networks, but also on energy distribution (closing of nuclear plants because water for cooling could be too hot or flooded plants, less hydroelectric power production because of droughts, etc).

The Internet, as we know it, ever more rapid and mobile, is the product of abundant and growing energy provision and a robust electricity infrastructure. It will not be able to survive if these conditions change. This is what Kris De Decker[13] shows in his article 'How to build a low-tech internet'. Consequently, the Internet server of the 'Low Tech Magazine' blog created by Kris De Decker functions on an architecture with a low-consumption processor supplied by a battery charged by solar panels. If the available energy is not sufficient, the blog is not accessible.

---

10 Moving follow the wind/follow the sun to production, Arnaud Greenstar, Available from: https://www.glif.is/meetings/2011/rap/arnaud-greenstar.pdf [Accessed 5 December 2018].

11 Arnaud St Bill, Preparing for climate change through energy internet and evehicles, 2009, Available from: https://green-broadband.blogspot.com/ [Accessed 5 December 2018].

12 In its preamble, this Canadian project cited a report by the UK government titled: How Climate Change Could Ruin the Internet.

13 Kris De Decker, 'How to build a low-tech internet', Low Tech Magazine, 2015, Available from: https://solar.lowtechmagazine.com/2015/10/how-to-build-a-low-tech-internet.html [Accessed 10 November 2017].

Conclusion

Fifty years after the first promises of cybernetics, the smart city has experienced many setbacks. Blocked on technological solutionism and urban reductionism,[1] the algorithmic strategies of the IT companies stumbled over the incommensurability and complexity of how cities function.[2] But a paradox persists: the end of the idealization of the smart city's urbanity has not slowed down the deployment of digital structures correlative to the acceleration of the world's digitalization and online life as far as the borders of the most rural zones in which it has become the right arm of the agriculture of tomorrow.[3] Whereas the anti-geographic dynamic of a virtual fallback behind computers increased even further with the Covid health crisis, the territories are continuing to equip themselves with a new infrastructural layer.[4] As for architecture, the balance of power seems unfavourable. If the excessiveness and technical proliferation trend has had a long life in the history of architecture, the digital has brought to its paroxysm what Reyner Banham[5] criticized with the technologies of the late 19th and early 20th centuries: the growth of the engineering of

1 Evgeny Morozov, *To Save Everything, Click Here: The Folly of Technological Solutionism*, New York, PublicAffairs, 2013; Ola Söderström, Till Paasche and Francisco Klauser, 'Smart cities as corporate storytelling', *City*, 18(3), 2014, pp 307–20; Rob Kitchin, 'The real-time city? Big data and smart urbanism', *GeoJournal*, 79, 2014, pp 1–14; Rob Kitchin, 'The promise and perils of smart cities', *Journal of the Society for Computers and Law*, 26(2), Available from: https://www.scl.org/3385-the-promise-and-perils-of-smart-cities/ [Accessed 26 November 2020].

2 Antoine Courmont, 'Où est passée la smart city? Firmes de l'économie numérique et gouvernement urbain', *Cities are back in town*, Working Paper no 02, Sciences Po, 2018, Available from: https://sciencespo.hal.science/hal-02186713/document

3 L'atelier Paysan, *Reprendre la terre aux machines*, Paris, Le Seuil, 2021.

4 Whereas the anti-geographic dynamic of a virtual fallback behind computers increased further with the Covid health crisis, the territories have continued to equip themselves with a new infrastructure layer; Fanny Lopez, *A bout de flux*, Paris, Divergences, 2022.

5 Reyner Banham, *L'Architecture de l'environnement bien tempéré*, Orléans, HYX, 2011 [original edn, *Architecture of the Well-Tempered Environment*, London, Architectural Press, 1969].

technical systems to the detriment of architecture considered a protective and durable envelope.[6]

There is no doubt that the impact of digital infrastructures on architecture, the city and the territories is comparable to the impact that the car, the train or electricity had on modern urbanism, but without the architecture that had accompanied them. In this large number of machines, the term 'architecture' is more commonly associated with the complexity of the network infrastructures' interconnections than with the millennial discipline that gives form to space. With the digital, technology could threaten – once again – architecture with disappearance.[7] This dualism identified by many historians of architecture and technologies has found a host of metaphorical descriptions, apprehending since the 1970s this continuum that seems to have lost any classic spatial reference. From the 'magical territory'[8] to the 'layer', by way of 'immaterial architecture' or the 'digital skin', these are images in which the interweaving of the material and the virtual triumph. Without architecture, technology once again takes the form of its simplest device: engines and motors, cogwheels, tubes, cables, etc, the whole covered by an envelope. If the spatiality of the digital uses several registers, it is close to an organization of technical systems arranged in boxes. The data centre is often drowned in the metropolitan fringes saturated by service spaces, notably electricity infrastructures, those 'terminal landscapes' identified by Jeff Diamanti.[9] This interconnection and storage structure complements the large family of decorated hangars alongside roads. In the globalized industry of data centre construction, architecture is no longer even a minimalist art to produce differentiation in the junk space. The lack of architecture perhaps mostly comes from the fact that the digital infrastructure is not a public program, but the result of a sum of private interests, in which architecture had often been associated with an expensive luxury. Architecture would henceforth be in the matrix. As announced in the 1990s, the architecture of network interfaces and Web design would take precedence over the territories of the real that is 'hardwarized' with data centres and power plants giving credit to the dystopic futurist visions of the first *Matrix* film (2000) and Edward Morgan Forster's short story *The Machine Stops*, written a century earlier (1909): a territorial reality loaded with infrastructures

6 Pierre Caye, *Durer- Éléments pour la transformation du système productif (Les Belles Lettres)*, 2020, pp 247–91.

7 Electricity replaces the millennial principle of bio-climatism (materials, orientation, etc) of electric heating.

8 Dominique Rouillard (ed), 'Territoire magique', in *Superarchitecture. Le futur de l'architecture 1950–1970*, Paris, Éditions de La Villette, pp 377–414.

9 Jeff Diamanti, *Climate and Capital in the Age of Petroleum: Locating Terminal Landscapes*, New York, Bloomsbury, 2021.

for a virtual life. With ever-more parts added or connected to the large number of machines, the cybernetics of the capitalist command seems to have been responsible for the disappearance of architecture, announced so many times. Data centres would be the masses and signs of this transfer, the advance of the cybernetic authority's gain over the territory and the literal loss of architecture. And this loss seems to be one of the major symptoms of the infrastructure crisis. Digital networks are developing in a post-public service age and isn't this the main reason for their lack of architecture? The infrastructures of the digital are proliferating and triumph in a world struck by the fall of public infrastructures.[10] To recover architecture, isn't it the common sense of the infrastructure that should be recaptured? Architecturing a general interest appears as a means of developing a public service and an architecture of the digital that is both fairer in its access and restrained in its electricity consumptions. What is imposed is the necessity of leaving behind the technological monism[11] and that of heavy public investment in order to give back form to the protective and liberating function of technology.[12] The electricity crisis of the digital infrastructure can be perceived as a historic opportunity to redefine our digital needs, to reorganize electricity distribution, envisage new energy pooling and build public architectures necessary for this new age of technology and other digital worlds.

10 Stephen J. Collier, James C. Mizes and Antina Von Schnitzler (eds), 'Public infrastructures/ infrastructural publics', *Limn*, 7, 2016, p 119.

11 Dominique Bourg and Christian Arnsperger, *Écologie intégrale, pour une société permacirculaire*, Paris, Presses Universitaires de France, 2017, p 150.

12 Caye, 'L'enveloppe protectrice'.

Bibliography

Oral sources

Interviews

Data centres in the Amsterdam region

Interview with Petra Boegheim, urbanist and Head of Data Centre Strategy, Amsterdam, March 2020.

Interview with Chantal Braak, Head of Grid Development, TenneT, Amsterdam, March 2020.

Interview with Ter Braak, regional development expert, TenneT, Amsterdam, March 2020.

Interview with Ruud Berndsen, Head of European Affairs, Alliander, March 2020.

Interview with Jeroen Graafland, Head of the Urban Planning Department, city of Haarlemmermeer, March 2020.

Interview with Irma Graveland, Administrative Communications Advisor/ Spokesperson for the Municipality of Haarlem, city of Haarlemmermeer, March 2020.

Interview with Jack Kranenburg, Commercial Director of Agriport A7, Agriport, March 2020.

Interview with Robert Kuik, Head of Network Development and Strategic Assets Management, TenneT, Amsterdam, March 2020.

Interview with Peter Kwakman, Head of Grid Development and Data Centres, TenneT, Amsterdam, March 2020.

Interview with Joost Molenaar, urbanism, public services strategy, utility strategy team, Amsterdam, March 2020.

Interview with Sander Oosterloo, network development expert, Alliander, Amsterdam, March 2020.

Interview with Roelof Potters, Head of Energy Transition, Alliander, Amsterdam, March 2020.

Interview with Ruban Rasing, Head of Grid Development, TenneT, Amsterdam, March 2020.

Interview with Bert Ross, Head of Relations and Environmental Strategy, Liander, Amsterdam, March 2020.

Interview with Tim Ruis, urbanist, regional development, sustainability, Head of Municipal Urban Heating Strategy, Amsterdam, March 2020.

Interview with Mariëtte Sedee, burgomaster, regional development, the environment and agricultural affairs, city of Haarlemmermeer, March 2020.

Interview with Clément Torre, urbanist, city of Haarlemmermeer, March 2020.

Interview with Ruben Voerman, urbanist, regional development and sustainability, Head of Municipal Electricity Strategy, Amsterdam, March 2020.

Data centres in Barcelona

Interview with Roger Baig of the Guifi.net Foundation, Barcelona, November 2018.

Data centres in California

Interview with Yen Han Chen, urbanist, Head of the Urbanism Department, city of Santa Clara, California, September 2017.

Interview with Richard Clark, Manager of the Vantage data centre campus, Santa Clara, California, September 2017.

Interview with Gary Cook, Head of the IT Department at Greenpeace, Editor-in-Chief of Reports on the Energy Impact of Digital Technologies, San Francisco, California, September 2017.

Interview with Anish Gautam, Engineer, California State Energy Commission, Sacramento, California, September 2017.

Interview with Fred Jalali, Energy Manager of the Zayo data centre in Santa Clara, California, September 2017.

Interview with Wendy Stone, Account Executive at Silicon Valley Power, municipal electricity company, Santa Clara, California, September 2017.

Data centres in Dublin

Interview with Noëlle Ameijenda, Production Analyst, EirGrid, February 2020.

Interview with Shane Brennan, specialist in regulations and contracts, EirGrid, February 2020.

Interview with Tom Gallery, Data Center Connection Manager, EirGrid, February 2020. Interview with Aisling Gilchrist, Engineer, EirGrid, February 2020.

Data centres in Île-de France

Interview with Benjamin Bayart, cofounder of Quadrature du Net, Co-President of the Fédération des Fournisseur d'Accès à Internet associatifs (FFDN), Paris, March 2018.

Interview with Régis Castagné, President of Equinix France, October 2020.

Interview with Fabrice Cocquio, President of Interxion France, Paris, March 2018 and September 2021.

Interview with Frederick Coeuille, General Manager/Director of Zayo France, Paris, June 2018.

Interview with Erik Ferrand, founder and President of Qarnot Computing, Paris, and Christophe Perron, founder and President of Stimergy, Paris, March 2018.

Interview with Guillaume Planchot, Heating and Cooling Network Director at Idex, March 2018.

Interview with Bruno Spiquel, activist at SCANI, Paris, July and November 2018.

Interview with Laurent Trescartes, cofounder and Director of Critical Building, Fontenay-aux-Roses, May 2018.

Interview with Patrick Vassallo, Community Delegate Advisor to the local development of Plaine Commune and Deputy Mayor for the Social and Solidarity Economy, Saint-Denis, February 2018.

Interview with Alain Vaucelle, Head of the ICT Mission for Plaine Commune, Saint-Denis, February 2018.

Data centres in New York

Interview with Ben Falber, Energy Storage Project Manager, New York State Energy Research and Development Authority, New York, January 2018.

Interview with David Schirmacher, President of 7×24 Exchange International, President of Reset Advisors, former Vice President of Operations for Digital Realty, New York, January 2018.

Interview with Stanley Voronov, Senior Manager, IBX Operations, north region, New York, January 2018.

Interview with the energy system managers of the Google data centre, New York, January 2018.

Data centres in Oregon

Interview with Crystal Ball, Head of Public Relations Bonneville Power Administration (BPA), Portland, Oregon, September 2017.

Interview with Matthew J. Chancellor, Regional Manager, PacifiCorp, Bend, Oregon, September 2017.

Interview with Mark Clemons, Director of Economic Development, Hillsboro, Oregon, September 2017.

Interview with Ted Reid, Head of Regional Urbanism, Portland Metro, Portland, Oregon, September 2017.

Interview with Letha Tawney, Innovation Director for Renewable Energies, World Resources Institute, Portland, Oregon, September 2017.

Interview with Paul Vaccaro, Vice President of Operations at Stack, Hillsboro, Oregon, October 2017.

Data centres in the Stockholm region

Interview with Henrik Bergström, Head of Public Affairs, Ellevio AB, videoconference, September 2020.

Interview with Catarina Grenemark, Head of Customer Relations on the regional network, videoconference, September 2020.

Interview with Hans Larsson, Electricity Systems Analyst, Vattenfall Eldistribution AB, videoconference, September 2020.

Interview with Stefan Lindbom, Commercial Project Manager for customer relations, videoconference, September 2020.

Interview with Örjan Lönngren, Policy Analyst for Energy and Climate at the city of Stockholm, October 2018.

Interview with Ulf Moberg, Technical Director, Svenska Kraftnät, videoconference, September 2020.

Interview with Erik Rylander, Head of the Stockholm Data Park Program, October 2018 and January 2022.

Interview with Johan Sjöborg, Network Analyst, videoconference, September 2020.

Interview with Klara Wirdby, Commercial Development Director at the municipality of Gävle, September 2020.

Printed sources

Data centres

Articles and chapters

Jenna Burrell, 'On half-built assemblages: waiting for a data center in Prineville, Oregon', *Engaging Science, Technology, and Society*, 6, 2020, pp 283–305.

Mél Hogan, 'Data flows and water woes : The Utah data center', *Big Data and Society*, 2(2), 2015, pp 1–12.

Mél Hogan, 'The data centre industrial complex', in M. Jue and R. Ruiz (eds), *Saturation: An Elemental Politics*, Durham, NC, Duke University Press, 2021, pp 283–305.

Bruno Moriset, 'Les forteresses de l'économie numérique. Des immeubles intelligents aux hôtels de télécommunications', *Géocarrefour*, 78(4), 2003, pp 375–88.

Jean-Christophe Plantin and Julia Velkova, 'Data centers and the infrastructural temporalities of digital media: An introduction', *New Media and Society*, 25(2), 2023, pp 273–86.

Asta Vonderau, 'Scaling the Cloud: making state and infrastructure in Sweden', *Ethnos*, 84(4), 2018, pp 698–718.

Books

Monika Dommann, Hannes Rickli and Max Stadler (eds), *Data Centers: Edges of a Wired Nation*, Zurich, Lars Müller, 2020.

Reports

Agence Anyoji and Beltrando, *Cahier de prescriptions urbaines, paysagères et architecturales, étude de programmation urbaine et économique*, December 2012. Archive du service de la ville.

Buck Consultants International, MRA-brede Strategie Data centers, CE Delft, 2020.

CE Delft, *Systeemstudie energie-infrastructuur Noord-Holland 2020–2050*, report, June 2019.

D'Cision, *Kansen en bedreigingen voor datacenters in de Metropool Regio Amsterdam*, 2019.

Jakob Dybdalx Christensen, Jens Therkelsen, Ivo Georgiev and Henrik Sand, *Data Center Opportunities in the Nordics: An Analysis of the Competitive Advantages*, report for the Nordic Council of Ministers, 2018.

Digital Gateway to Europe (Stijn Grove, Judith de Lange), *North Amsterdam Data Center Campus: Economic Impact Study*, Report, June 2018, Available from: http://www.digitalgateway.eu/uploads/1/6/1/8/16187644/northamsterdamdccampus_eng.pdf [Accessed 28 February 2025].

Cécile Diguet and Fanny Lopez, *L'impact spatial et énergétique des data centers sur les territoires*, ADEME report, 2019.

French Ministry of Industry, the Caisse des Dépôts et Consignations, and the Territorial Equity Agency (eds) *Guide on Cloud Computing and Data Centers for Local Authorities*, 2015.

Host and Bitpower, *Ireland's Data Hosting Industry 2019 Q3 Report*, Available from: https://static.wixstatic.com/ugd/914803_6e99282e6ce54b66ae04533408904c24.pdf [Accessed 28 February 2025].

Irish Academy of Engineering, 'Electricity Sector Investment for Data Centers in Ireland', note, July 2019.

Ruimtelijk economische ontwikkel strategie, *Ruimtelijke Strategie Datacenters Routekaart 2030 voor de groei van datacenters in Nederland*, March 2019.

Theses and HDR

Clément Marquet, *Binaire béton: Quand les infrastructures numériques aménagent la ville*, under the direction of Jérôme Denis, thesis defended on 30 October 2019 at Université Paris-Saclay (ComUE), in the framework of the École doctorale Sciences de l'Homme et de la société in partnership with Télécom Paris (Palaiseau) (thesis preparation establishment) and the Institut interdisciplinaire de l'innovation (Paris) (laboratory).

The city and the digital territory

Articles and chapters

Antoine Courmont, 'Où est passée la smart city? Firmes de l'économie numérique et gouvernement urbain', *Cities are back in town*, Working Paper no 02, Sciences Po, 2018, Available from: https://sciencespo.hal.science/hal-02186713/document

Orit Halpern, 'L'architecture comme machine: la ville intelligente déconstruite', in Andrew Goodhouse (ed), *Quand le numérique marque-t-il l'architecture?*, London, Stenberg Press, 2017, pp 121–76.

Rob Kitchin, 'The real-time city? Big data and smart urbanism', *GeoJournal*, 79(1), 2014, pp 1–14.

Rob Kitchin, 'The promise and perils of smart cities', *Journal of the Society for Computers and Law*, 26(2), Available from: http://www.scl.org/site.aspx?i=ed42789 [Accessed 28 November 2020].

Rem Koolhaas, 'Brave world: people, places, technology', *Flaunt*, 145, 2016, p 145.

Pierre Musso, 'Critique de la notion de "territoires numériques"', *Quaderni*, 66, spring 2008, pp 15–29.

Ola Söderström, Till Paasche and Francisco Klauser, 'Smart cities as corporate storytelling', *City: Analysis of Urban Trends, Culture, Theory, Policy, Action*, 18(3), 2014, pp 307–20.

Valérie Peugeot, 'Collaborative ou intelligente? La ville entre deux imaginaires', in Maryse Carmes and Jean-Max Noyerp (eds), *Devenirs urbains*, Paris, Presses des Mines, 2016, pp 43–64.

Books

Antoine Picon, *Smart cities: Théorie et critique d'un idéal auto-réalisateur*, Paris, B2, 2013.

Antoine Picon, *Smart Cities – A Spatialised Intelligence*, Chichester, Wiley, 2015.

Anthony M. Townsend, *Smart Cities, Big Data, Civic Hackers, and the Quest for a New Utopia*, New York, W. W. Norton & Co, 2013.

HDR

Soline Nivet, Architecture et *numérique: Infrastructures, Interfaces & Institutions d'un 'écosystème' parisien*, HDR, Université Paris-Est Sup, 15 December 2020.

Reports

Michele Bebrin, Kenneth Horne and Jonathan Strahl (eds), *Community Microgrid Case Study and Analysis Report*, report for the New York State Smart Grid Consortium, New York, Navigant Consulting, 2015.

Jean Daniélou with François Ménard, *L'art d'augmenter les villes (pour) une enquête sur la ville intelligente*, La Défense, Plan Urbanisme construction architecture, 2013, Available from: http://www.urbanisme-puca.gouv.fr/l-art-d-augmenter-les-villes-pour-une-enquete-sur-a342.html [Accessed 12 November 2020].

Evgeny Morozov and Francesca Bria, *Rethinking the Smart City, Democratizing Urban Technology*, New York, Rosa Luxemburg Stiftung, 2018.

The digital, energy and the environment

Articles and chapters

Anders S. G. Andrae and Edler Tomas, 'On global electricity usage of communication technology: trends to 2030', *Challenges*, 6, 2015, pp 117–57.

Françoise Berthoud and Laurent Lefevre, 'Impact écologique du BigData', in Mokrane Bouzeghoub and Rémy Mosseri (eds), *Les Big Data à découvert*, Paris, CNRS, 2017, pp 84–9.

Françoise Berthoud, Amélie Bohas, Laurent Lefevre and Marianne Parry, 'Transformation numérique et transformation énergétique – Digital transformation and energy transformation', in Agathe Euzen, Stéphanie Thiébault and Bettina Laville (eds), *Quelles solutions pour le changement climatique? Which solutions for climate change?*, Paris, CNRS, 2015, pp 67–83.

Kris De Decker, 'How to build a low-tech internet', *Low Tech Magazine*, 2015, Available from: https://solar.lowtechmagazine.com/2015/10/how-to-build-a-low-tech-internet.html [Accessed 10 November 2017].

Rolf Landauer, 'Irreversibility and heat generation in the computing process', *IBM Journal of Research and Development*, 5(3), 1961, pp 183–91.

Anne-Cécile Orgerie and Laurent Lefèvre, 'Energy-efficient data transfers in large-scale distributed systems', in Ishfaq Ahmad and Sanjay Ranka (eds), *Handbook of Energy-Aware and Green Computing*, New York, Chapman & Hall, 2012, pp 965–82.

Books

Anil Agarwal and Sunita Narain, *Global Warming in an Unequal World: A Case of Environmental Colonialism*, New Delhi, Centre for Science and Environment, 1991.

François Deltour, Michelle Dobré, Fabrice Flipo and Marion Michot, *Peut-on croire aux TIC vertes? Technologies de l'information et crise environnementale*, Paris, Presses des Mines, 2012.

Paul N. Edwards, *Computer Models, Climate Data, and the Politics of Global Warming*, Cambridge, MA, MIT Press, 2013.

Fabrice Flipo, *La face cachée du numérique*, Paris, Éditions de L'Échappée, 2013.

Fabrice Flipo, *L'impératif de la sobriété numérique: L'enjeu des modes de vie*, Paris, Matériologiques, 2020.

Fabrice Flipo, *Le numérique, une catastrophe écologique*, Paris, L'Échappée, 2021.
Groupe EcoInfo, *Impacts écologiques des Technologies de l'Information et de la Communication – Les faces cachées de l'immatérialité*, Les Ulis, EDP Sciences, 2012.
Guillaume Pitron, *L'enfer numérique: Voyage au bout d'un Like*, Paris, Liens qui libèrent, 2021.

Reports

Association négaWatt, *Scénario négaWatt 2017–2050*, 2017, p 8, Available from: https://www.negawatt.org/IMG/pdf/synthese_scenario-negawatt_2017-2050.pdf [Accessed 2 March 2018].
City of Amsterdam and Liander, *Thematische studie elektriciteit Amsterdam*, March 2019.
Deloitte Développement Durable, Eco Info, Futuribles and the CRÉDOC, *Potentiel de contribution du numérique à la réduction des impacts environnementaux. État des lieux et enjeux pour la prospective*, ADEME report, 2016.
Hugues Ferreboeuf (ed), *Lean ICT, Pour une sobriété numérique*, The Shift Project report, October 2018.
Mohan Ganeshalingam, Arman Shehabi and Louis-Benoit Desroches, *Shining a Light on Small Data Centers in the U.S.*, Energy Analysis and Environmental Impacts Division, Lawrence Berkeley National Laboratory, June 2017.
Région Île-de-France, Schéma Régional du Climat, de l'Air et de l'Energie de l'Île-de-France, RTE, October 2018, Paris, Available from: https://www.drieat.ile-de-france.developpement-durable.gouv.fr/IMG/pdf/14-SRCAE-IDF_Synthese_actions_recommandees_collectivites_cle4114e6_cle18d814.pdf [Accessed 14 March 2025].

The energy landscape

Books

Jeff Diamanti, *Climate and Capital in the Age of Petroleum: Locating Terminal Landscapes*, New York, Bloomsbury Publishing, 2021.

Reports

Thomas Veyrenc, Executive Director of RTE and Xavier Piechaczyk, Chairman of the Board of RTE, RTE press conference, 'Futurs énergétiques 2050', Paris, 25 October 2021.

Technology, territory, society

Articles and chapters

Stephen J. Collier, James C. Mizes and Antina Von Schnitzler (eds), 'Public infrastructures/infrastructural publics', *Limn*, 7, 2016.
Gabriel Dupuy, 'Fracture et dépendance: l'enfer des réseaux?', *Flux*, 1(83), 2011, pp 6–23.

Stephen Graham, 'Global grids of glass: on global cities, telecommunications and planetary urban networks', *Urban Studies*, 36, 1999, pp 929–49.

Cédric Le Bris and Olivier Coutard, 'Les réseaux rattrapés par l'environnement? Développement durable et transformations de l'organisation des services urbains', *Flux*, 4(74), 2008, pp 6–10.

Fanny Lopez, '*Keep the lights on!* La décentralisation énergétique à Londres', in Gilles Lepesant (ed), *Énergies nouvelles, territoires autonomes?*, Paris, Presses de l'Inalco, 2018, pp 27–60.

Jonathan Rutherford, 'Networks in cities, cities in networks: territory and globalisation intertwined in telecommunications infrastructure development in Europe', *Urban Studies*, 42(13), 2005, pp 2389–406.

Jonathan Rutherford, 'Le déploiement des réseaux de télécommunications dans les métropoles européennes: l'emprise des territorialités Multiscalaires', *Métropolis Flux*, 75, 2009, pp 6–15.

Books

Wiebe E. Bijker, Thomas Parke Hughes and Trevor Pinch (eds), *The Social Construction of Technological Systems*, Cambridge MA, MIT Press, 1984.

Sarah Brayne, *Predict and Surveil: Data, Discretion, and the Future of Policing*, New York, Oxford University Press, 2021.

Manuel Castells, *La Société en réseaux. L'ére de l'information*, Paris, Fayard, 1998.

Benjamin Coriat, Nicole Alix, Jean-Louis Bancel and Frédéric Sultan (eds), *Vers une république des biens communs*, Paris, Les liens qui libèrent, 2018.

Gabriel Dupuy, *Aux origines des sciences cognitives*, Paris, La Découverte, 1994.

Gabriel Dupuy, *Internet: Géographie d'un Réseau*, Paris, Ellipses, 2002.

Jacques Ellul, *Le Système technicien*, Paris, Calmann-Lévy, 1977.

Ali Fard and Taraneh Meshkani (eds), *Geographies of Information*, New Geographies 07, Harvard University Graduate School of Design, October 2015.

Stephen Graham and Simon Marvin, *Splintering Urbanism: Networked Infrastructures, Technological Mobilities and the Urban Condition*, London, Routledge, 2001.

Alain Gras, *Grandeur et dépendance, sociologie des macro-systèmes techniques*, Paris, Presses Universitaires de France, 2013.

Steve J. Heims, *Constructing a Social Science for Postwar America*, Cambridge, MA, MIT Press, 1991.

Thomas Parke Hughes, *Networks of Power: Electrification in Western Society, 1880–1930*, Baltimore, MD, Johns Hopkins University Press, 1983.

Thomas Parke Hughes and Renate Mayntz, *The Development of Large Technical Systems*, Frankfurt, Campus, 1988.

Fanny Lopez, *Le rêve d'une déconnexion, de la maison autonome à la cité auto-énergétique*, Paris, La Villette, 2014, translation *Disconnection Dream, from the Autonomous House to Self-Sufficient Territories*, Manchester University Press, 2021.

Fanny Lopez, *L'ordre électrique, infrastructures énergétiques et territoires*, Geneva, Métis Presses, 2019.

L'atelier Paysan, *Reprendre la terre aux machines*, Paris, Le Seuil, 2021.

Jonathan Rutherford, *A Tale of Two Global Cities: Comparing the Territorialities of Telecommunications Developments in Paris and London*, Aldershot, Ashgate, 2004.

Theses

François Huguet, (Re)coudre avec du sans fil. Enquête sur des pratiques de médiation infrastructurelle, thesis under the direction of Annie Gentès and Jérôme Denis, Paris ENST, defended in June 2016. Notably see the summary of the thesis, Available from: https://medium.com/@francoishuguet/re-coudre-avec-du-sans-fil-ae0d4f28888c [Accessed 26 February 2025].

Internet, cybernetics and the digital culture

Articles and chapters

Roger Baig, Felix Freitag and Leandro Navarro, Cloudy in guifi.net: establishing and sustaining a community cloud as open commons, ScienceDirect, Available from: https://www.sciencedirect.com/science/article/pii/S0167739X1732856X [Accessed 26 February 2025].

Roger Baig, Lluís Dalmau, Ramon Roca, Leandro Navarro, Felix Freitag and Arjuna Sathiaseelan, Making Community Networks Economically Sustainable: The Guifi.net Experience, in Proceeding GAIA 16, Proceedings of the 2016 workshop on Global Access to the Internet, 2016, Available from: http://people.ac.upc.edu/leandro/pubs/baig-sigcomm.pdf [Accessed 26 February 2025].

Primavera de Filipi and Félix Tréguer, 'Les réseaux Mesh', in Cécile Méadel and Francesca Musiani (eds), *Abécédaire des architectures distribuées*, Paris, Presses des Mines, Collection I, 2015.

Primavera de Filippi and Félix Tréguer, 'Expanding the Internet commons: the subversive potential of the Wireless Community Network', *Journal of Peer Production*, 6, Disruption and the Law, 2015, Available from: https://halshs.archives-ouvertes.fr/halshs-01306630v2/document [Accessed 26 February 2025].

Francesca Musiani and Valérie Schafer, 'Le modèle internet en question (1970–2010)', *Flux*, 3(85–6), 2011, pp 62–71.

Francesca Musiani, 'Stockage distribué', Cécile Méadel and Francesca Musiani, *Abécédaire des architectures distribuées*, Paris, Presses des Mines, 2015, p 209.

Books

Andrew Blum, *Tubes: A Journey to the Center of the Internet*, New York, Harper Collins, 2012.

Jay Forrester, *Urban Dynamics*, Cambridge, MA, MIT Press, 1969.

Cécile Méadel and Francesca Musiani, *Abécédaire des architectures distribuées*, Paris, Presses des Mines, 2015.

Donella Meadows, Dennis Meadows and Jørgen Randers, *Beyond the Limits*, White River Junction, VT, Chelsea Green Publishing, 1992.

Evgeny Morozov, *To Save Everything, Click Here: The Folly of Technological Solutionism*, New York, PublicAffairs, 2013.

Nicholas Negroponte, *The Architecture Machine*, Cambridge, MA, MIT Press, 1970.

Nicholas Negroponte, *The Architecture Machine: Toward a More Human Environment*, Cambridge, MA, MIT Press, 1973.

Félix Tréguer, *L'Utopie déchue, une contre-histoire d'internet, XVe–XXIe siècle*, Paris, Fayard, 2019.

Fred Turner, *From Counterculture to Cyberculture: Stewart Brand, the Whole Earth Network, and the Rise of Digital Utopianism*, Chicago, University of Chicago Press, 2012.

Norbert Wiener, *Cybernetics or Control and Communication in the Animal and the Machine,* Cambridge MA, MIT Press, 2019.

The history and philosophy of techniques

Articles

Joerges Bernward, 'Large technical systems: concepts and issues', in T. P. Hugues and R. Mayntz (eds), *The Development of Large Technical Systems*, Frankfurt, Campus, 1998, pp 9–32.

Books

Pierre Caye, *Critique de la destruction créatrice. Production et humanisme*, Paris, Les Belles Lettres, 2015.

Pierre Caye, *Durer: éléments pour la transformation du système productif*, Paris, Les Belles Lettres, 2020.

Lewis Mumford, *Technics and Civilization*, New York, Harcourt, Brace & Co, 1934.

Gilbert Simondon, *Du mode d'existence des objets techniques* (1989), Paris, Aubier, 2012.

Alain Supiot, *La gouvernance par les nombres, cours ou Collège de France (2012–14)*, Paris, Fayard, 2015.

Alexis de Tocqueville, *De la démocratie en Amérique II* (1840), chapter XII, Paris, Gallimard, 1961.

Architecture

Articles and chapters

Fanny Lopez, 'Les fantômes des télécoms: l'obsolescence des grands centraux public', in Dominique Rouillard (ed), *Public – Infrastructure, architecture, territoire*, Paris, Beaux-Arts, 2021, pp 175–89.

Dominique Rouillard, 'Le monument, la démocratie, l'objet', *Cahiers du CCI*, 3, 1985, pp 25–8.

Books

Maurice Culot (ed), *Architectures de l'électricité. Architectures de l'âge industriel,* Paris, Norma, 1992.

Andrew Goodhouse (ed), *Quand le numérique marque-t-il l'architecture?*, London, Stenberg Press, 2017.

Dean Hawkes, *The Environmental Tradition: Studies in the Architecture of Environment*, Oxford, Taylor & Francis, 1996.

Marc Levinson, *How the Shipping Container Made the World Smaller and the World Economy Bigger*, Princeton NJ, Princeton University Press, 2008.

Dominique Rouillard, *Superarchitecture, le futur de l'architecture 1950–1970,* Paris, La Villette, 2004.

Dominique Rouillard and Alain Guiheux, *Door to door: futur du véhicule, futur urbain*, Paris, Archibooks, 2016.

Dominique Rouillard and Alain Guiheux, *Le projet Hub – L'architecture des nouvelles mobilités*, Geneva, Métis Presses, 2020.

Ecology and energy

Books

Christian Arnsperger and Dominique Bourg, *Écologie intégrale, pour une société permacirculaire*, Paris, Presses Universitaires de France, 2017.

William Behrens, Donnella Meadows, Dennis Meadows and Jorgen Randers, *The Limits to Growth*, New York, Universe Books, 1972.

Fanny Lopez, *A bout de Flux*, Paris, Éditions Divergences, 2022.

Jason W. Moore (ed) *Anthropocene or Capitalocene?: Nature, History, and the Crisis of Capitalism*, Kairos, PM Press, 2016.

The economy, the digital economy

Articles and chapters

Antoine Courmont, 'Où est passée la smart city? Firmes de l'économie numérique et gouvernement urbain', *Cities are back in town,* Working Paper no 02, Sciences Po, 2018, Available from: https://sciencespo.hal.science/hal-02186713/document

Books

Digital Realty, *Data Gravity Index DGx, Report measuring the intensity of Data Gravity and its effect on the Global 2000 Enterprises*, Digital Realty, 2020.

Tim Jackson, *Prosperity without Growth: Foundations for the Economy of Tomorrow*, London and New York, Earthscan/Routledge, 2009.

Yann Moulier-Boutang, *L'Abeille et l'Économiste*, Paris, Carnets Nord, 2010.

E. Ostrom, *Governing the Commons: The Evolution of Institutions for Collective Action*. Cambridge, Cambridge University Press, 1990.

Daniel Schiller, *Digital Capitalism: Networking the Global Market System*, London, MIT Press, 1999.

Pierre Veltz, *Mondialisation, villes et territoires: Une économie d'archipel*, Paris, Presses Universitaire de France, 1996.

Index

References to figures appear in *italic* type. References to endnotes show both the page number and the note number (126n12).

A

activists 10, 150–2
administration (enterprise) data centres 28, 93n10
Adobe data centre *59*
Agriport (Netherlands) 82–4
alternative digital structures 10, 133–4
  associative and cooperative ISPs 144–6
  data centres of proximity 146–50
  and digital sovereignty 135–7
  ecological footprint 162–5
  future scenario 172–3
  peer-to-peer model 138–43
  for resilience and inclusion 155–61
  self-management in Catalonia 150–2
Amazon Web Services (AWS) 1, *25*, 34, 79–81
Amitié cable 92n8
Amsterdam
  data centre locations 49–52, 65–6
  digital infrastructure strategy 120, 121, 126–7
  electricity supply 122–6
Amsterdam Data Tower 51, *52*
Anarchaserver project 151–2, *153*
anonymity *see* discretion/anonymity
Apple data centre 76–8, 100
Aquilenet 147–9
architecture
  data centre characteristics 7–8, 23–8, 55–6, 78, 79–80
  and digital infrastructure 178–9
  see also internet architecture
Arduinna laboratory 152
Arpanet 19
Ashburn (Virginia) 93
associative data centres 146–50
associative ISPs 144–6
AT&T 38
AtNorth 129
Avenue of Americas (New York) 38, *40*, *42*, *43*, *44*, *45*
AWS *see* Amazon Web Services

B

back-up energy supply 100–1, 108, 109
  for community networks 158
  fuel oil generators 109–10, 114–15
  gas cogeneration 110–12, 113–14
BATX 170
Bayart, Benjamin 145
Big Tech companies (GAFAM) 86–90, 99–100, 169, 170–1
  see also specific companies
Black Hills Energy 114
boulevard Voltaire (Paris) 48, *49*
Bria, Francesca 17
Brista data park 129–30
Brodie, Patrick 5
Burrell, Jenna 5

C

cables
  in New York 37–8, *40*, *41*, *42*
  see also optical fibre
Calafou cooperative 150–2
Canada 177
Catalonia 150–2
Céleste data centre 117, *118*
centralization
  of digital infrastructure 19–22
  of digital storage 33–4
  future scenario 169–72
  of Internet architecture 169–70
Les Chatons collective 146
Childs, Quincy 5
cities *see* metropolitan data centres
citizen networks *see* community/citizen networks
citizen opposition 12, 70, 88

climate change 177
climate events 109, 139, 157–8, 175
Cloud solutions 135–7, 142–3
Cloudy 143
clusters 58–9, 61–5, 107, 126
CO^2 emissions 172, 175, 176
cognitive capitalism 16
colocation data centres 28, 30, 34, 50, 68–9, 93–4, 93n10
commons 141n7, 152–4, 165
communication networks (mesh) 138–42, 156, 158
communities
  advantages of pooling for 117–19, 135–6
  Big Tech contributions to 78, 87
  future control by 174
  resilience and inclusion 155–62
  see also citizen opposition
community/citizen networks 152–4
  and associative/cooperative data centres 146–50
  associative/cooperative ISPs 144–6
  Cloud solutions 143
  mesh networks 138, 139–40
  Red Hook Initiative 155–8
  self-management in Catalonia 150–2
  W4C initiative 160
company (enterprise) data centres 28, 93n10
computer equipment 19, 163
computer simulation 16
concentration of data 91, 93–4
confidentiality 101
Cook, Gary 99
cooperative data centres 146–50
cooperative ISPs 144–6
cooperatives, Calafou 150–2
Coresite data centre *57*
La Courneuve (France) 69, 70–3, *74*
crisis, response to 13–14, 15
cybernetics 14–15
cyberspace, utopian vision of 17
CyrusOne 66

D

The Dalles (US) *24*, *25*
Daniélou, Jean 16
dark optical fibre 55, 127n14
Data Centre Parks program (Stockholm) 127–31
data centres
  administration/enterprise 28, 93n10
  architectural characteristics 7–8, 23–8, 55–6, 78, 79–80, 147
  associative/cooperative 146–50
  colocation 28, 30, 34, 50, 68–9,93–4, 93n10
  concentration of 93–4
  electricity connection in Île-de-France 103–8
  emergence and growth 20–1, 33
  and energy *see* back-up energy supply; electricity consumption; electricity destabilization
  equipment *42*, *43*, *44*, *45*
  equipment sharing 114–15
  future diversity of 173–4
  heat recovery 105, 115–19, 128, 129–30
  interdependence 34
  local concerns 1, 36, 64, 67–8, 70, 88
  local support 78–9, 80–1, 87
  location *see* location of data centres
  regulation *see* regulation
  scholarship on 4–6
  size 29–30, 78
data centres of proximity 146–50, 162–4
data mules 160–1, 176
data storage
  distributed 142–3
  emergence of 20–1
  industry trends 33–4
  see also data centres
De Decker, Kris 177
de Filippi, Primavera 138–9, 143
decentralized and distributed infrastructures
  associative ISPs 144–6
  data centres of proximity 146–50
  data mules 160–1, 176
  digital commons 152–4
  early computers 19
  ecological footprint 162–5
  future scenarios 172–7
  Red Hook Initiative 155–8
  self-management in Catalonia 150–2
  sneakernet 161
  W4C initiative 160
  see also peer-to-peer model
Delay Tolerant Networks (DTN) 160–1
Deux Fleurs association 146
digital activists 10, 150–2
digital growth 2, 91–3, 169–72
digital industry 2, 33–6
  in Sweden 85–6
  see also Big Tech companies
digital infrastructure strategy 131–2
  in Netherlands 120–7
  in Sweden 127–31
digital infrastructures
  and architecture 178–9
  centralization of 19–22
  discrete 21–2
  and electricity destabilization 99–102
  energy consumption of 1, 2, 97–9
  energy efficiency of 97–8
  growth of 21
  spatial evolution 34–5
  see also alternative digital structures; decentralized and distributed infrastructures; digital technical system; peer-to-peer model
Digital Realty data centres 51

digital sovereignty 10, 135–7, 150–2
digital technical system (DTS)
  alternatives *see* alternative digital structures
  decentralization 175–7
  future growth and centralization 169–72
  future stabilization 172–5
  invisibility of 21–2
  scholarship on 3–6
  see also data centres; digital infrastructures; internet infrastructure
discretion/anonymity
  of buildings 7–8, 55–6, 78, 79–80
  of digital infrastructures 21–2
Disruption Tolerant Networks (DTN) 160–1
distributed communication 138–42
distributed infrastructure *see* decentralized and distributed infrastructures
distributed storage 142–3, 162–4
diversity, in future scenarios 173–4
Dobigny, Laure 163
Dommann, Monika 5
DRIEE (Regional and Interdepartmental Division of Environment and Energy) 105
DTN (Delay/Disruption Tolerant Networks) 160–1
DTS *see* digital technical system
Dublin 61–5
Dunan cable 92n7

E

e-commerce 170
ecological impacts
  of alternative digital structures 162–5
  see also climate events
EDF (France) 103n1
8th avenue (Chelsea, New York) 44–6, *47*
electricity consumption
  and data centre size 29–30
  of data centres of proximity 149
  of digital infrastructures 1, 2, 97–9
  Dutch scenarios 122
  flexible demand measures 64–5
  Île-de-France 105–6, 107
  increased 97–9, 107
  rural data centres 78–9
electricity costs, and location 54, 58, 61
electricity destabilization
  impact of data centres 64, 70, 80–1, 96, 99–102, 121–2
  impact of digital infrastructure 99–102
  regulatory response to 67, 131
electricity grids 16–17, 100–2, 105–7
  see also micro-energy grids; substations
electricity production
  destabilization of 99–100
  renewable sources *see* renewable energies
electricity supply
  back-up *see* back-up energy supply
  connections in Île-de-France 103–8
  and data centre location 30–1
  destabilization *see* electricity destabilization
  Dutch scenarios 122–7
  reservation of 65, 101, 106, 121
  substations 62–4, 83–4, 88, 89, 105, 106, 121–6
  see also substations
Ellalink cable 92
employment 60, 78
Enedis 103n1, 105–6
energy
  digital infrastructure impacts 1–2, 8–9
  electrification of 107
  future scenarios 171–2, 175, 177
  reservation of 65, 101, 106, 121
  strategy in Netherlands 121–7
  see also back-up energy supply; *electricity headings*; fuel oil generators; gas; micro-energy grids; pooling; renewable energies
energy consumption
  of alternative digital structures 163
  of data centres *see* electricity consumption
  of digital infrastructures 1, 2, 97–9
  future scenarios 172, 175, 177
energy efficiency 97–8
  see also pooling
enterprise (administration) data centres 28, 93n10
Equinix data centres 50–1, *57*, 69
equipment *see* computer equipment; data centres
Ersbo Microsoft data centre 88
Europe
  data centre locations 35–6
  see also specific countries

F

Facebook cables 92n6, 92n8
Facebook data centres 76–8, 86, 87n20
FDN (French Data Network) 144–5
feminist servers 151–2
Feng, Wu 97
fibre optic cables *see* optical fibre
flexible demand 64–5
flooding 157–8
Forrester, Jay 15
Fortum 127, 129
fossil fuels 121, 127
  see also fuel oil generators; gas
France
  associative ISPs 144–6
  data centre energy in Île-de-France 103–8
  data centre locations 31, 47–9, 68–75
  data centres of proximity 146–50

digital industry growth 2
digital sovereignty 135–7
gas cogeneration 111–12
heat recovery 105, 115–19
Franciliens.net 145
Free Software movement 19
French Data Network (FDN) 144–5
fuel cells 113
fuel oil generators 109–10, 114–15
Fuller, Richard Buckminster 14
future digital worlds
growth and centralization 169–72
stabilization and diversity 172–5
ultra-decentralization 175–7

G

GAFAM
future scenarios 170–1
influence of 169
role in electricity destabilization 99–100
in Sweden 86–90
see also specific companies
gas, used at Grange Castle 64
gas cogeneration 83, 110–12, 113–14
gas production 121
Gävle (Sweden) 86–90
Google cables 92n7
Google data centres
The Dalles *24*, *25*
Dublin *64*
Netherlands 81–2, 83, *85*
New York 44–6
governance 170
see also regulation
Grange Castle Business Park 62–5
Green Star project 177
greenhouses 84, *84*
growth
concerns about 15–16
future digital growth 169–72
role of optical fibre 91–3
Guifi.net 141–2, 143

H

Haarlemmermeer data centres 66–8, 121
Haché, Alexandra 151–2
Halpern, Orit 13
heat recovery 105, 115–19, 126n12, 127–31, 171
Hillsboro digital zone 58–61
Hollands Kroon (Netherlands) 81–5
housing 70, 119, 121
Hudson Street (New York) 38–9
Huguet, François 140
Hurricane Sandy 109, 157
hydroelectricity 58, 86, 100
hyperscale data centres 29–30, 31, 34, 62, 82–4, 86–90

I

Idex data centre 117
Île-de-France 1, 103–8
see also Paris
Iliad data centres 49, 117–19
inclusion initiatives 155–61
India 160
industrial sites 26, 37–47, 48–9, 53, 54, 58, 151
infrastructure data centres *see* colocation data centres
interdependence, of data centres 34
Intergate Manhattan Tower 39–41
intermittent connectivity 160–1, 175–6
Internet architecture, future scenarios 169–70, 172–3, 175–7
Internet Exchange Points (IXPs) 20, 26, 30, 35, 47, 54
Internet infrastructure
centralized 19–22
decentralized 155–61, 175–7
distributed 138–42
see also community/citizen networks
Internet Service Providers (ISPs) 144–6
Interxion data centres 69, 70–3, *74*, 117, 129, 164
Ireland 61–5
IT refresh processes 163
IXPs *see* Internet Exchange Points

K

Kista data park 128–9, 131
Koomey, Jonathan 97

L

Landauer, Rolf 97n2
large technical systems (LTS) 170
Lehuedé, Sebastián 5
location of data centres 26, 30–1, 35, 131–2
data centres of proximity 147, 149
Dutch strategy 120–1
Swedish strategy 127, 131
see also metropolitan centres; rural data centres; urban outskirts data centres
Loudoun County (Virginia) 26, 35

M

Maréa cable 92n6
Marseille 101
Ménard, François 16
mesh networks 138–42, 156, 158
metropolitan data centres 31, 35, 37
Amsterdam 49–52
future scenarios 171
New York 37–47
Paris 47–9
micro-data centres 29, 35
micro-energy grids 83, 100–1, 113–14, 158, 172

micro-scale futures 176
Microsoft
  colocation partners 34
  data centres *63*, 82–3, 86, 87–90, 100
  energy pooling 114
Moriset, Bruno 26, 30
Morozov, Evgeny 17
Moulier-Boutang, Yann 16
Musiani, Francesca 142

N

NAPs (network access points) 20
Natixis bank data centre 117
natural disasters 109, 139, 157–8, 175
Negroponte, Nicholas 15
Netherlands
  data centre locations 49–52, 65–8, 81–5
  digital infrastructure strategy 120–7
network access points (NAPs) 20
network interconnection points *see* Internet Exchange Points
New York
  data centre back-up systems 109–10, 113–14
  data centre locations 37–47
  Red Hook Initiative 155–8
New York Stock Exchange 37
New York Telephone Company 40
The New York Times building 114
Nivet, Soline 35

O

oil *see* fuel oil generators
Open Source movement 19
optical fibre
  dark fibre 55, 127n14
  and data centre location 30, 48
  and digital growth 91–3
  emergence of 19–20
Orange data centres *48*, 117
Oregon *see* The Dalles; Hillsboro digital zone; Prineville
Ostrom, Elinor 141
OVH data centre *59*

P

Palo Alto 54
Paris 47–9, 117
Peace cable 91
peer-to-peer (P2P) model 138–43, 174
personal computers 19
Plaine Commune 68–75, 93–4
Plantin, Jean-Christophe 5
pooling
  of equipment/services 114–15, 135–6
  heat recovery 116–19, 126, 127–30
Portland, Hillsboro digital zone 58–61
Portland Dispatchable Standby Generators 114–15
Power Usage Effectiveness (PUE) 9
Prineville (Oregon) 76–81
privatization 19, 20

Q

Qarnot (company) 119

R

reactive environment 15
Red Hook Initiative (RHI) 155–8
regulation 35
  construction denied 1, 67, 131
  and digital sovereignty 135–7
  of electricity supply 106
  future scenarios 173, 175
  in Netherlands 67–8, 106, 126–7
renewable energies
  future scenarios 171–2, 177
  and GAFAM companies 81, 86, 99–100
  and intermittent networks 161
  in Ireland 100
  in Netherlands 121
  in Sweden 86, 128
  in United States 81, 100
  and user behaviour 163
REOS report 120
reserved energy 65, 101, 106, 121
resilience 155–61
Rickli, Hannes 5
RightMesh 139
RTE (France) 103n1, 105, 106–7
rural data centres 31, 76, 171
  Netherlands 81–5
  Sweden 86–90
  United States 76–81
Rylander, Erik 130

S

Sabey data centre 41, 110
San José (US) 56, *57*
Sandviken (Sweden) 86–90
Santa Clara (US) 54–8
SCANI 146–7
Schiphol (Netherlands) 66
Sciences Park (Amsterdam) 49–52
security
  and centralization 20
  data centre characteristics 26, 78
  of digital infrastructures 164
  digital sovereignty and 135–7
  and peer-to-peer models 143
self-hosting 142, 143
self-management 150–2
SFR 49
Sidewalk Labs 12, 170
Silicon Valley 31, 54–8
Silicon Valley Power (SVP) 54–5
siting *see* location of data centres

Skarpnäck data park 130
SLTE (Submarine Line Terminal Equipment) 92–3
smart city concept 6, 11–12
 antecedents 14–15
 limitations 16–17
 vision of 13–14
sneakernet networks 161
solar energy 100
South Dublin County Council (SDCC) 62–4
sovereignty 10, 135–7, 150–2
spatial impact of alternative digital structures 163–4
spatial impact of digital infrastructures 1–2
Spelman, Caroline 177
Spideralex 151–2
Spiquel, Bruno 147
'splintering urbanism' 171
SRCAE plan 105
Stack data centre *59*, 60–1
Stackbo Microsoft data centre 88, 89
Stadler, Max 5
Stimergy 119
Stockholm 127–31
Stockholm Exergi 127, 128
Stokab 127n14
Submarine Line Terminal Equipment (SLTE) 92–3
substations 62–4, 83–4, 88, 89, 105, 106, 121–6
Supiot, Alain 12
SVP (Silicon Valley Power) 54–5
Sweden 85–90, 115, 127–31
system dynamics 15

T

tax incentives 59, 60, 61, 66, 78, 86n17
technological sovereignty 10, 135–7, 150–2
Telecity data centre 117
telecom buildings, conversion of 37–44, 46–7, 147
telecom companies 38–9, 40–1, 44–5
Telehouse operator 48, *49*
temperature control 27, 58, 110, 114, 117
 see also heat recovery
Tetraneutral 149–50
Tocqueville, Alexis de 21n5
Toronto 12
transportation, data mules 161
Tréguer, Félix 17, 138–9
Tuna Microsoft data centre 88, 89
Turner, Fred 17–18
2Africa cable 91–2

U

Umatilla (Oregon) 79–81
undersea cables 37–8, 91–2
United States
 back-up energy supplies 109–10, 113–15
 data centre locations *24*, *25*, 26, 35, 37–47, 54–61, 76–81, 93
 Red Hook Initiative 155–8
Université de Bourgogne 119
urban centres *see* metropolitan data centres
urban districts
 future scenario 170–1
 Stockholm data centre parks 127–32
urban outskirts data centres 31, 53
 Dublin 61–5
 Hillsboro 58–61
 Netherlands 65–8
 Plaine Commune 68–75
 Santa Clara 54–8

V

Vantage campus (Santa Clara) 56
Väsby data park 130
Velkova, Julia 5
Verizon 41

W

W4C (Wireless for Communities) 160
wavelengths 92–3
Western Union 38–9
Wi-Fi networks 138–40, 176
Wiener, Norbert 14
wind-power 100
Wireless for Communities (W4C) 160
World Game (Fuller) 14

Z

Zayo data centres 31, 48